Viral Genes and Plant Pathogenesis

Thomas P. Pirone John G. Shaw
Editors

Viral Genes and Plant Pathogenesis

With 50 Illustrations

Springer-Verlag
New York Berlin Heidelberg
London Paris Tokyo Hong Kong

Thomas P. Pirone
Department of Plant Pathology
University of Kentucky
Lexington, KY 40546-0091
USA

John G. Shaw
Department of Plant Pathology
University of Kentucky
Lexington, KY 40546-0091
USA

Library of Congress Cataloging-in-Publication Data
Viral genes and plant pathogenesis / Thomas P. Pirone, John G. Shaw, eds.
p. cm.
Papers presented at the Symposium on Viral Genes and Plant Pathogenesis, held at Lexington, Ky., October 16 and 17, 1989.
1. Plant viruses—Genetics—Congresses. 2. Virus diseases of plants—Genetic aspects—Congresses. I. Pirone, T. P. II. Shaw, John G. III. Symposium on Viral Genes and Plant Pathogenesis (1989: Lexington, Ky.)
QR351.V5 1990
581.2′34—dc20 90-9744

Printed on acid-free paper.

Camera-ready copy provided by the editors.
Printed and bound by Edwards Brothers, Ann Arbor, Michigan.
Printed in the United States of America.

9 8 7 6 5 4 3 2 1

ISBN 0-387-97313-3 Springer-Verlag New York Berlin Heidelberg
ISBN 3-540-97313-3 Springer-Verlag Berlin Heidelberg New York

Preface

The chapters in this book represent detailed versions of papers presented at the Symposium on Viral Genes and Plant Pathogenesis held at Lexington, Kentucky on October 16 and 17, 1989. In selecting topics and authors, we attempted to have represented a spectrum of systems which are at the forefront of research on plant virus genes and gene products, particularly as they relate to plant disease. The book also contains pertinent discussion of the papers presented at the symposium, as well as summaries, observations and projections of future research directions prepared by the session chairmen. We wish to express our appreciation to Dr. D.L. Davis, of the R.J. Reynolds Tobacco Company for suggesting the organization of the Symposium and the publication of the proceedings, and to the R. J. Reynolds Tobacco Company for the financial support which made the symposium possible. We also wish to thank those of our colleagues in the Department of Plant Pathology, University of Kentucky, who very ably and in many ways contributed to the organization and conduct of the conference.

Thomas P. Pirone
John G. Shaw

Contents

Contributors

Paul Ahlquist
Institute for Molecular Virology
Department of Plant Pathology
University of Wisconsin
Madison, WI 53706

Richard Allison
Institute for Molecular Virology
Department of Plant Pathology
University of Wisconsin
Madison, WI 53706

J.G. Atabekov
Department of Virology
Moscow State University
Moscow 119899
USSR

Roger N. Beachy
Department of Biology
Washington University
St. Louis, MO 63130

David M. Bisaro
The Biotechnology Center
Department of Molecular Genetics
The Ohio State University
Columbus, OH 43210

J.F. Bol
Gorlaeus Laboratories
Leiden University
Einsteinweg 5
2333 CC Leiden
THE NETHERLANDS

J.-M. Bonneville
Friedrich-Miescher-Institut
P.O. Box 2543
CH4002 Basel
SWITZERLAND

S. Bouzoubaa
Institut de Biologie Moléculaires des Plantes
12 Rue de Général Zimmer
67084 Strasbourg Cédex
FRANCE

Clare L. Brough
The Biotechnology Center
Department of Molecular Genetics
The Ohio State University
Columbus, OH 43210

G. Bruening
CSIRO Division of Plant Industry
GPO Box 1600
Canberra ACT 2601
AUSTRALIA

B.J.C. Cornelissen
MOGEN Int.
Einsteinweg 97
2333 CB Leiden
THE NETHERLANDS

William O. Dawson
Department of Plant Pathology and
Genetics Graduate Group
University of California
Riverside, CA 92521

Walter Dejong
Institute for Molecular Virology
Department of Plant Pathology
University of Wisconsin
Madison, WI 53706

M. de Tapia
Friedrich-Miescher-Institut
P.O. Box 2543
CH4002 Basel
SWITZERLAND

William G. Dougherty
Department of Microbiology
Oregon State University
Corvallis, OR 97331

J. Fütterer
Friedrich-Miescher-Institut
P.O. Box 2543
CH4002 Basel
SWITZERLAND

W.L. Gerlach
CSIRO Division of Plant Industry
GPO Box 1600
Canberra ACT 2601
AUSTRALIA

Rob Goldbach
Department of Virology
Wageningen Agric. University
THE NETHERLANDS

Karen-Beth Goldberg
Department of Plant Pathology
University of Kentucky
Lexington, KY 40546

K. Gordon
Friedrich-Miescher-Institut
P.O. Box 2543
CH4002 Basel
SWITZERLAND

Siddarame Gowda
Department of Plant Pathology
University of Kentucky
Lexington, KY 40546

H. Guilley
Institut de Biologie Moléculaires des Plantes
12 Rue de Général Zimmer
67084 Strasbourg Cédex
FRANCE

Rosemarie W. Hammond
Microbiology and Plant Pathology Laboratory
Plant Sciences Institute
USDA-ARS
Beltsville, MD 20705

B.D. Harrison
Scottish Crop Research Institute
Invergowrie
Dundee DD2 5DA
UNITED KINDGOM

Marcos Hartitz
The Biotechnology Center
Department of Molecular Genetics
The Ohio State University
Columbus, OH 43210

J.P. Haseloff
CSIRO Division of Plant Industry
GPO Box 1600
Canberra ACT 2601
AUSTRALIA

T. Hohn
Friedrich-Miescher-Institut
P.O. Box 2543
CH4002 Basel
SWITZERLAND

Sheriar G. Hormuzdi
The Biotechnology Center
Department of Molecular Genetics
The Ohio State University
Columbus, OH 43210

Roger Hull
Virus Research Department
John Innes Institute
Norwich NR4 7UH
UNITED KINGDOM

A.G. Hunt
Department of Agronomy
University of Kentucky
Lexington, KY 40546

Michael Janda
Institute for Molecular Virology
Department of Plant Pathology
University of Wisconsin
Madison, WI 53706

J. Jiricny
Friedrich-Miescher-Institut
P.O. Box 2543
CH4002 Basel
SWITZERLAND

G. Jonard
Institut de Biologie Moléculaires des Plantes
12 Rue de Général Zimmer
67084 Strasbourg Cédex
FRANCE

Isabelle Jupin
Institut de Biologie Moléculaires des Plantes
12 Rue de Général Zimmer
67084 Strasbourg Cédex
FRANCE

S. Karlsson
Friedrich-Miescher-Institut
P.O. Box 2543
CH4002 Basel
SWITZERLAND

Jennifer Kiernan
Department of Plant Pathology
University of Kentucky
Lexington, KY 40546

O.A. Kondakova
Department of Virology
Moscow State University
Moscow 119899
USSR

Philip Kroner
Institute for Molecular Virology
Department of Plant Pathology
University of Wisconsin
Madison, WI 53706

John A. Lindbo
Department of Microbiology
Oregon State University
Corvallis, OR 97331

H.J.M. Linthorst
Gorlaeus Laboratories
Leiden University
Einsteinweg 5
2333 CC Leiden
THE NETHERLANDS

S.I. Malyshenko
Department of Virology
Moscow State University
Moscow 119899
USSR

Tetsuo Meshi
Department of Biophysics and Biochemistry
Faculty of Science
University of Tokyo
Hongo, Tokyo
JAPAN

A.R. Mushegian
Department of Virology
Moscow State University
Moscow 119899
USSR

Ursula Niesbach-Klösgen
Institut de Biologie Moléculaires des Plantes
12 Rue de Général Zimmer
67084 Strasbourg Cédex
FRANCE

Yoshimi Okada
Department of Biosciences
School of Science and Engineering
Teikyo University
Utsunomiya, Tochigi
JAPAN

Robert A. Owens
Microbiology and Plant Pathology Laboratory
Plant Sciences Institute
USDA-ARS
Beltsville, MD 20705

Radiya Pacha
Institute for Molecular Virology
Department of Plant Pathology
University of Wisconsin
Madison, WI 53706

Peter Palukaitis
Department of Plant Pathology
Cornell University
Ithaca, NY 14853

T. Dawn Parks
Department of Microbiology
Oregon State University
Corvallis, OR 97331

T.P. Pirone
Department of Plant Pathology
University of Kentucky
Lexington, KY 40546

L. Quillet
Institut de Biologie Moléculaires des Plantes
12 Rue de Général Zimmer
67084 Strasbourg Cédex
FRANCE

Gwen N. Revington
Department of Botany and Microbiology
Auburn University
Auburn, AL 36849

R.E. Rhoads
Department of Biochemistry
University of Kentucky
Lexington, KY 40546

K. Richards
Institut de Biologie Moléculaires des Plantes
12 Rue de Général Zimmer
67084 Strasbourg Cédex
FRANCE

H. Sanfaon
Friedrich-Miescher-Institut
P.O. Box 2543
CH4002 Basel
SWITZERLAND

J.E. Schoelz
Department of Plant Pathology
University of Kentucky
Lexington, KY 40546

H. Scholthof
Department of Plant Pathology
University of Kentucky
Lexington, KY 40546

M. Schultze
Friedrich-Miescher-Institut
P.O. Box 2543
CH4002 Basel
SWITZERLAND

J.G. Shaw
Department of Plant Pathology
University of Kentucky
Lexington, KY 40546

R.J. Shepherd
Department of Plant Pathology
University of Kentucky
Lexington, KY 40546

Michael Shintaku
Department of Plant Pathology
Cornell University
Ithaca, NY 14853

Holly A. Smith
Department of Microbiology
Oregon State University
Corvallis, OR 97331

Garry Sunter
The Biotechnology Center
Department of Molecular Genetics
The Ohio State University
Columbus, OH 43210

M.E. Taliansky
Department of Virology
Moscow State University
Moscow 119899
USSR

Patricia Traynor
Institute for Molecular Virology
Department of Plant Pathology
University of Wisconsin
Madison, WI 53706

C.M.A. van Rossum
Gorlaeus Laboratories
Leiden University
Einsteinweg 5
2333 CC Leiden
THE NETHERLANDS

Yuichiro Watanabe
Department of Biophysics and Biochemistry
Faculty of Science
University of Tokyo
Hongo, Tokyo
JAPAN

Fang C. Wu
Department of Plant Pathology
University of Kentucky
Lexington, KY 40546

M.J. Young
CSIRO Division of Plant Industry
GPO Box 1600
Canberra ACT 2601
AUSTRALIA

Milton Zaitlin
Department of Plant Pathology
Cornell University
Ithaca, NY 14853

Induction of Host Genes by the Hypersensitive Response of Tobacco to Virus Infection

J.F. Bol, C.M.A. van Rossum, B.J.C. Cornelissen[1] and H.J.M. Linthorst

Department of Chemistry, Gorlaeus Laboratories, Leiden University, Einsteinweg 5, 2333 CC Leiden and [1]MOGEN Int., Einsteinweg 97, 2333 CB Leiden, The Netherlands

As far as is known, a mild systemic infection of a plant by a virus has little effect on the pattern of host gene expression. The activity of an enzyme with properties *in vitro* of an RNA-dependent RNA-polymerase is enhanced by such an infection in many plant species (Fraenkel-Conrat, 1983). In addition, cDNA has been cloned to a 0.9 kb host mRNA that is induced by the compatible interaction of cucumber with cucumber mosaic virus (Linthorst *et al.*, 1989a). On the other hand, dozens of host genes are known to be induced by the hypersensitive response of plants to infection with a necrotizing virus. These genes appear to be involved in various defence reactions resulting in a direct attack of the invading pathogen, a localization of the pathogen at the site of infection and an adaptation of the host metabolism to the stress condition evoked by the infection (for a review see Bol *et al.*, 1990). A group of stress-induced proteins that have been characterized in detail are the so-called "pathogenesis-related" (PR) proteins. Initially, these were defined as acidic, protease-resistant proteins that accumulated in the intercellular space of tobacco leaves responding hypersensitively to tobacco mosaic virus (TMV) infection (for a review see Van Loon, 1985). More recently, it became clear that basic isoforms of the majority of acidic PR proteins accumulate in the vacuoles of cells in TMV-infected tobacco leaves. These basic isoforms have also been identified in many other plant species reacting hypersensitively to infection with viruses, fungi and bacteria, or as a result of treatment of the plant with numerous biotic or abiotic elicitors (see Van Loon, 1985; Bol, 1988; Bol *et al.*, 1990). As far as they have been analysed, corresponding proteins of different plant species were reported to have an amino acid sequence identity of approximately 60%, indicating that PR genes are highly conserved in the plant kingdom. Over 20 acidic and basic PR proteins have been recognized in TMV-infected tobacco. On the the basis of their molecular properties and serological relationships, these proteins have been classified into five different groups (Jamet and Fritig, 1986; Van Loon *et al.*, 1987; Kauffmann, 1988; Fritig *et al.*, 1989). These groups are briefly discussed below.

Group 1 contains the PR-1 proteins of which the function is not known. The genome of Samsun NN tobacco contains about eight genes for acidic PR-1 proteins and a similar number of genes encoding basic PR-1 proteins (Cornelissen *et al.*, 1987). After TMV-infection three genes encoding acidic

PR-1 proteins (PR-1a, -1b and 1c) and at least one of the genes encoding basic PR-1 proteins are expressed. The mol. wt. of the acidic PR's is about 14.5 kd; a serologically related 16 kd protein, found in TMV-infected tobacco, could represent their basic isoform (Kauffmann, 1988). The sequence of a cDNA clone indicates that the basic PR-1 protein is synthesized as a 19 kd precursor with a C–terminal extension of 36 amino acids (Cornelissen *et al.*, 1987). Genomic clones of a number of PR–1 genes have been isolated and sequenced by several research groups (see Bol *et al.*, 1990).

Group 2 contains acidic and basic isoforms of enzymes with ß-1,3-glucanase activity. The acidic isozymes are represented by PR proteins 2 (39.7 kd), N (40.0 kd), 0 (40.6 kd) and Q' (36 kd) whereas a single basic ß-1, 3-glucanase (Gluc. b, 33.0 kd) was identified (Kauffman *et al.*, 1987; Fritig *et al.*, 1989). These proteins are related by serology (Fritig *et al.*, 1989) and amino acid sequence (Van den Bulcke *et al.*, 1989; Linthorst *et al.*, unpublished). Genomic blots indicate that the acidic and basic isozymes are each encoded by families of four to eight genes (Shinshi *et al.*, 1988; Linthorst *et al.*, unpublished).

Group 3 contains two acidic chitinases, the PR proteins P (27.5 kd) and Q (28.5 kd), and two basic chitinases, named Ch. 32 (32.0 kd) and Ch. 34 (34.0 kd) (Legrand *et al.*, 1987). These four enzymes are related by serology and amino acid sequence. It has been estimated that the Samsun NN genome contains two to four genes for both acidic and basic isozymes (Hooft van Huijsduijnen *et al.*, 1987). Chitin, the substrate of chitinases, is not present in healthy plants but is found among others in the cell wall of fungi. The combination of chitinase and ß-1,3-glucanase has been shown to be highly effective in inhibiting the growth of a number of fungi *in vitro* (Mauch *et al.*, 1988) and may be responsible for the TMV-induced resistance of tobacco to fungal infection (Gianinazzi, 1983).

Group 4 contains a class of low molecular weight proteins of unknown function. One protein of 13 kd and another one of 15 kd were considered to be isoforms of a PR protein called R by Van Loon *et al.*, (1987) and R' by Pierpoint (1986). Recently Kauffmann (1988) and Fritig *et al.*, (1989) recognized two 13.0 kd proteins (named r2 and s2) and two 14.5 kd proteins (named r1 and s1). These proteins have not yet been characterized in detail.

Group 5 consists of several acidic and basic proteins with a molecular weight of 24 kd. Initially, one acidic protein was identified which was called PR-S by Van Loon *et al.*, (1987) and PR-R by Pierpoint (1986). Later on, two acidic 24 kd proteins were recognized, named R-minor and R-major by Pierpoint *et al.*, (1987) and R and S by Fritig *et al.*, (1989). The genes encoding these two acidic 23 kd proteins have been cloned (Van Kan *et al.*, 1989). In addition to the acidic 24 kd proteins, at least one basic 24 kd protein is induced by TMV-infection (Fritig *et al.*, 1989) that is probably identical to a salt-stress induced tobacco protein, called osmotin (Sing *et al.*, 1987). Both the acidic and basic 24 kd tobacco proteins show an approximately 65% amino acid sequence identity to the sweet-tasting protein thaumatin (Cornelissen *et al.*, 1986) and a maize inhibitor of α-amylase and protease of insects (Richardson *et al.*, 1987). By analogy, the tobacco 24 kd proteins could be involved in an anti-insect defence.

In addition to PR proteins from groups 1 to 5 a number of other host proteins are known to be induced by viral infection. These include enzymes involved in the biosynthesis of aromatic compounds, enzymes involved in lignification, cell wall proteins, etc. (see Bol *et al.*, 1990). We have characterized a family of genes encoding glycine-rich proteins (GRP) which are highly induced by TMV-infection of tobacco (Hooft van Huijsduijnen *et al.*, 1986; Van Kan *et al.*, 1988).

One of our approaches to obtain insight in the putative role of PR proteins in defence mechanisms is to express the corresponding genes constitutively in transgenic plants. Plants expressing PR genes in the sense orientation are analysed for a possible increased resistance to pathogens while plants expressing antisense transcripts are assayed for the opposite effect. Moreover, these plants could be useful in the identification of the signal molecule(s) responsible for the systemic induction of PR genes at a distance from the site of infection. It has been shown that PR proteins initially accumulate around single local lesions of TMV infected tobacco (Antoniw and White, 1986). The possibility that hydrolytic enzymes from groups 2 and 3 release oligosaccharide signal molecules from damaged plant cell walls could be tested by an inhibition of the induction of the corresponding genes by an antisense strategy. Here we report on the susceptibility to insect attack of transgenic plants expressing genes encoding PR-1a (group 1), PR-S (group 5) and GRP in the sense orientation. To assess the feasibility of the anti-sense strategy, the induction of PR-S in transgenic plants accumulating PR-S antisense transcripts was investigated.

EXPRESSION OF SENSE PR TRANSCRIPTS IN TRANSGENIC PLANTS

Previously, we reported that tobacco plants transformed with PR-1, PR-S or GRP genes under the transcriptional control of the cauliflower mosaic virus 35S promoter, selectively accumulated the corresponding mRNAs or proteins at levels comparable to those in TMV-infected plants (Linthorst *et al.*, 1989b). In that study the nomenclature for group 5 PR proteins of Van Loon *et al.*, (1987) was used. Although the gene used in the transformation experiments probably corresponds to R-minor and R in the nomenclatures of Pierpoint *et al.*, (1987) and Fritig *et al.*, (1989), respectively, we shall adhere to the name PR-S in the present study.

Spraying of tobacco plants with 5 mM salicylate results in the rapid induction of a number of PR genes, notably genes encoding acidic and basic PR-1 genes, acidic ß-1,3-glucanases, basic chitinases and GRP (Hooft van Huijsduijnen *et al.*, 1986b; Bol *et al.*, 1989). Moreover, this treatment reduces the susceptibility of the plants to systemic infection with alfalfa mosaic virus (AlMV) by 90% (Hooft van Huijsduijnen *et al.*, 1986a). Because PR-1a and GRP are most strongly induced by salicylate treatment, these proteins could function in an antiviral response. However, the transgenic plants expressing

PR-1a, GRP and PR-S were as susceptible to infection with AlMV or TMV as were non-transformed control plants (Linthorst *et al.*, 1989b).

In view of the amino acid sequence similarity between PR-S and a maize inhibitor of digestive enzymes of insects, it was of interest to assay the transgenic plants for a possible resistance to insect attack. Larvae of the beet armyworm (*Spodoptera exigua*), (eggs kindly provided by Dr. J. Vlak, University of Wageningen, NL) and the tobacco budworm (*Heliothis virescens*), (eggs kindly provided by A. Stoker, Duphar, Weesp, NL) were reared according to Patana (1985). Insects in the larval stadia L1 to L5 were fed in petri dishes on detached leaves from either transgenic (PR-1a, GRP or PR-S) or non-transformed tobacco plants. Testing of larval stadia L1 to L5 of the beet armyworm did not show any difference between transgenic and control plants (Linthorst *et al.*, 1989a). Only larval stadia L1, L2 and L3 of the tobacco budworm could be used because the cannibalistic behaviour of this insect prevented the testing of L4 and L5 stadia. Fig. 1 shows leaves from a non-transformed tobacco plant and a PR-S transformed plant after four days of feeding by the L2 larval stadium of the tobacco budworm. No difference in susceptibility to this insect was observed between transgenic plants expressing PR-1a, GRP or PR-S and non-transformed control plants. Recently, resistance to the tobacco budworm has been reported for transgenic tobacco plants expressing a cowpea trypsin inhibitor (Hilder *et al.*, 1987).

EXPRESSION OF ANTI-SENSE PR TRANSCRIPTS IN TRANSGENIC PLANTS

A PR-S cDNA clone (pROB12) was engineered in such a way that the coding sequence was flanked by BamHI restriction sites at both ends (Linthorst *et al.*, 1989). This coding sequence, containing nucleotides +31 to +885 of PR-S mRNA, was inserted in the BamHI site of the transformation vector pROK 1 (Baulcombe *et al.*, 1986) behind the CaMV 35S promoter. Constructs with the cDNA in the reverse orientation were selected and transferred from *Eschericia coli* to *Agrobacterium tumefaciens* strain LBA4404 (Hoekema *et al.*, 1983) by triparental mating using plasmid pRK2013 in *E.coli* strain HB101 (Ditta *et al.*, 1980). Leaf discs of Samsun NN tobacco plants were infected with *A.tumefaciens* and kanamycin-resistant plants were regenerated from transformed shoots as described by Horsch *et al.*, (1985).

Expression of the chimeric gene was analysed in F1-progeny plants by Northern blot hybridization. Strand-specific probes were prepared by reverse priming of single stranded M13 DNA containing PRS cDNA in two different orientations (Hu and Messing 1982). Nine independent transformants, named E-minus plants 1 to 9, were analysed. The Northern blots of Fig. 2 show the accumulation of PR-S minus-strand RNA (panel A) and plus-strand RNA (panel B) in a number of E-minus plants and healthy (lane H) and TMV-infected (lane I) non-transformed control plants. Fig. 2A shows that the E-minus plants accumulate various levels of PR-S minus-strand RNA while no

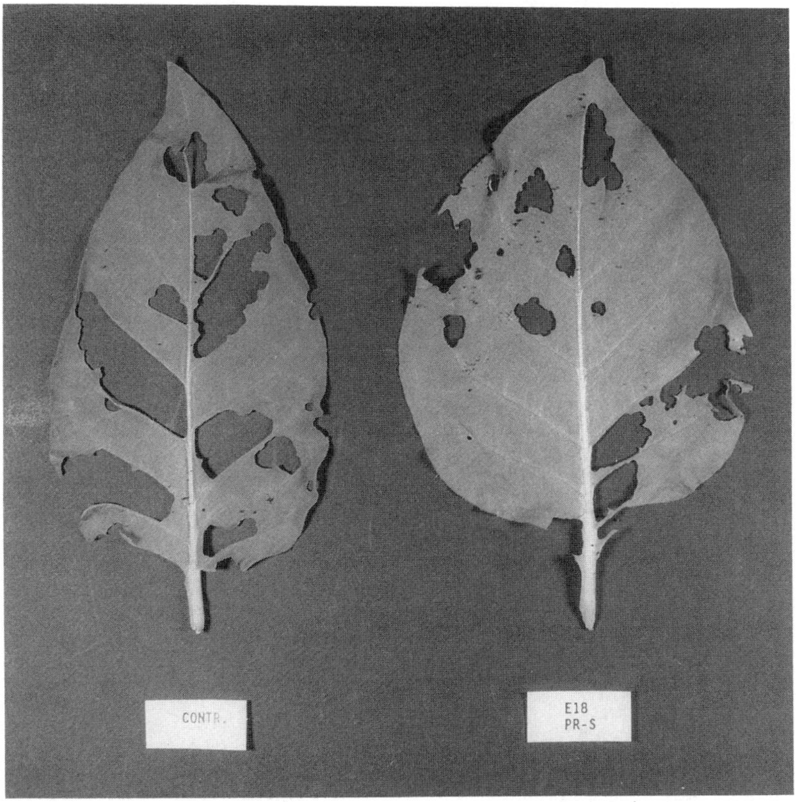

Fig. 1 Susceptibility of transgenic tobacco plants to insect attack. Leaves from a non-transformed control plant (left) and transgenic plant E18 expressing the PR-S gene in the sense orientation (right) were fed for four days to the L2 larval stadium of the tobacco budworm (*Heliothis virescens*).

minus-strand RNA is detectable in the controls. Fig. 2B shows that in the TMV-infected control plants the endogenous PR-R/S genes are efficiently induced but no PR-S plus-strand RNA is detectable in the healthy E-minus plants. Moreover, when this blot was hybridized to other PR cDNAs no signal was obtained (result not shown). This indicates that in these E-minus plants there is no induction of endogenous PR genes by the transformation procedure. The transgenic plants accumulating PR-S minus-strand RNA looked phenotypically normal.

E-minus plants were inoculated with TMV and 2, 5 and 8 days after inoculation protein extracts from the inoculated leaves were analysed by Western blotting. The blots were incubated with a mixture of antisera to PR proteins P/Q (Hooft van Huijsduijnen *et al.*, 1987) and R/S (Linthorst *et al.*, 1989). Fig. 3 shows that substantial amounts of PR proteins P/Q and R/S were detectable in nontransformed control plants, 5 or 8 days after infection with TMV. A similar accumulation of PR proteins P/Q was detectable in TMV-infected E-minus plants but compared to the controls the accumulation of PR proteins R/S in E-minus plants 1 and 9 was reduced by at least 80%. Fig. 3 shows the results obtained with plant 9; a similar reduction of R/S was observed in plant 1, but in plants 2, 4 and 5 the reduction was much lower (results not shown). The expression of PR-S anti-sense RNA did not affect the formation of TMV lesions.

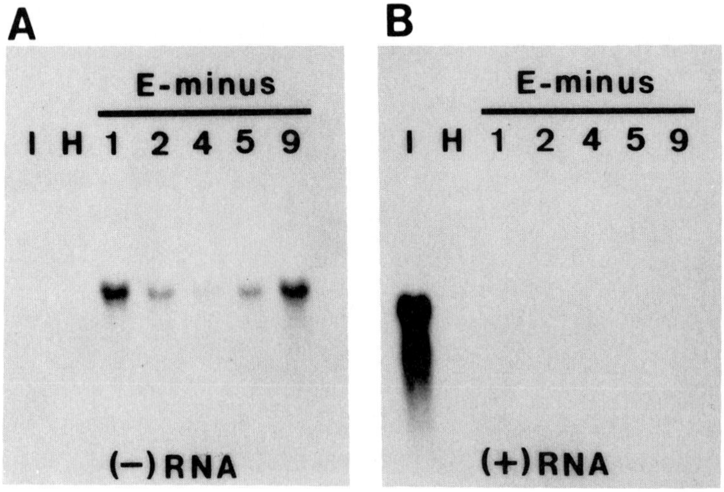

Fig. 2 Accumulation of PR-S antisense RNA in transgenic plants. Northern blots were loaded with RNA from transgenic plants expressing the PR-S gene in the antisense orientation (E-minus plants 1, 2, 4, 5 and 9), and RNA from non-transformed healthy (lanes H- and TMV-infected (lanes I) tobacco plants. The blots were hybridized to strand-specific probes detecting either PR-S minus-strand RNA (panel A) or PR-S plus-strand RNA (panel B).

In systemically induced leaves of TMV-infected non-transformed tobacco plants the accumulation of PR proteins R/S is dectable from about eight days after inoculation. In systemically induced leaves of E-minus plants 1 and 9 the reduction in the accumulation of PR-R/S was similar to that observed in the primary inoculated leaves while there was no effect on the accumulation of PR-P/Q (result not shown).

CONCLUSIONS

After TMV-infection of Samsun NN tobacco, many dozens of host proteins are induced that could act either separately or in concert in the various defence reactions collectively known as induced resistance (Bol *et al.*, 1990). The results obtained so far with transgenic plants expressing these proteins constitutively, indicate that PR-1, GRP and PR-S alone are not responsible for the TMV- and salicylate-induced resistance to virus infection or for a possible defence of tobacco to insect attack. The observation that most of the virus-inducible host proteins are encoded by multi-gene families may complicate the identification of (a combination of) host proteins that play an active role in induced resistance.

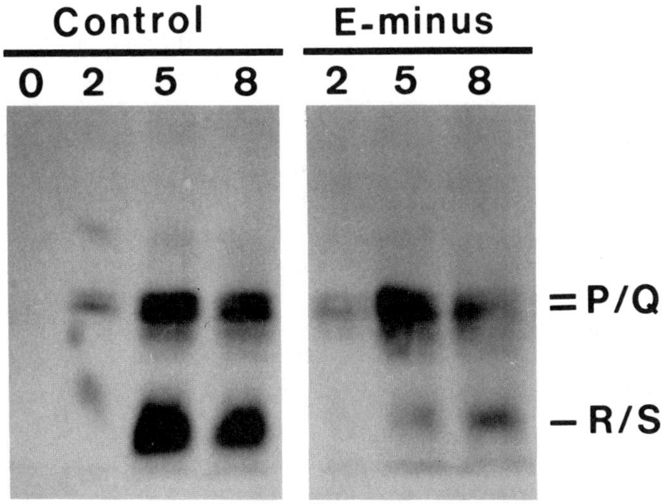

Fig. 3 Accumulation of PR proteins in E-minus plants after TMV-infection. Non-transformed control plants and E-minus plants were inoculated with TMV and 2, 5 and 8 days after infection protein extracts were made from the inoculated leaves and analysed by Western blotting. The blots were incubated with a mixture of antisera to PR proteins P/Q and R/S; the position of these proteins is indicated in the right margin. Figures on top of the lanes give the number of days after inoculation that protein was extracted.

In plant systems, the synthesis of anti-sense RNA has been shown to inhibit the expression of e.g. chalcone synthase genes (Van der Krol *et al.*, 1988) and polygalacturonase genes (Smith *et al.*, 1988) to levels similar to those obtained in this study with PR-S antisense RNA. In general, a high ratio of antisense to sense RNA seems to be required for effective reduction of the target gene and a

100% inhibition may never be reached. It may depend on the function of the target gene whether or not an 80 to 90% inhibition is sufficient to result in a physiological effect. The nucleotide sequence similarity between the genes encoding the two acidic 24 kd proteins from the group 5 of P/R proteins is over 90% (Van Kan *et al.*, 1988) and probably PR-S antisense RNA will inhibit both genes.

The sequence of the gene(s) encoding the basic 24 kd protein, called osmotin, is not yet known and the inhibition of this TMV-inducible gene by PR-S anti-sense RNA may be much less effective. The basic 24 kd protein has been shown to be serologically related to the acidic PR proteins R/S (Kauffmann, 1988) and it may well be that the residual amount of protein detected with the PR-S antiserum in TMV-infected E-minus plants is in fact the basic 24 kd protein. Similarly, the expression of more than one anti-sense gene may be required to inhibit the induction by TMV infection of acidic and basic isoforms of PR proteins from groups 1, 2 and 3. Anyhow, the observation that a large reduction in the level of PR-R/S in E-minus plants does not affect TMV lesion formation indicates that these proteins have no role in the localization of the infection. Moreover, the finding that other PR proteins are systemically induced in TMV-infected E-minus plants suggests that PR-R/S are not essential for the formation of the signal molecule that is responsible for the induction of PR genes in non- inoculated leaves. The anti-sense strategy will be useful to analyse the putative role of other PR proteins in the localization of a pathogen at the site of infection and the generation of long distance signal molecules. In addition, this strategy could shed light on the function of the root-specific and flower-specific expression in healthy plants that has been observed for a number of PR proteins (Memelink *et al.*, 1989; Lotan *et al.*, 1989).

REFERENCES

Antoniw JF, White RF (1986) Changes with time in the distribution of virus and PR protein around single local lesions of TMV infected tobacco. *Plant Mol Biol* **6**: 145-149.

Baulcombe DC, Saunders GR, Bevan MW, Mayo MA, Harrison BD (1986) Expression of biologically active viral satellite RNA from the nuclear genome of transformed plants. *Nature* (London) **321:** 446- 449.

Bol JF (1988) Structure and expression of plant genes encoding pathogenesis-related proteins. In: Verma DPS, Goldberg RB (eds) Plant Gene Research; Temporal and Spatial Regulation of Plant Genes, pp. 201-221. Springer Verlag, Wien/New York.

Bol JF, Linthorst HMJ, Cornelissen BJC (1990) Plant pathogenesis-related proteins. In: Gould H (ed) Annual Review of Phytopathology, Vol 28. Annual Reviews Inc., Palo Alto, in press.

Bol JF, Van de Rhee MD, Van Kan JAL, González Jaén MT, Linthorst HJM (1989) Characterization of two virus- inducible plant promoters. In: Lugtenberg BJJ (ed) Molecular Signals in Microbe-Plant Symbiotic and Pathogenic Systems. Springer Verlag, Wien, in press.

Cornelissen BJC, Hooft van Huijsduijnen RAM, Bol JF (1986) A tobacco mosaic virus-induced tobacco protein is homologous to the sweet tasting protein thaumatin. *Nature* **321**: 531- 532.

Cornelissen BJC, Horowitz J, Van Kan JAL, Goldberg RB, Bol JF (1987) Structure of tobacco genes encoding pathogenesis-related proteins from the PR-1 group. *Nucleic Acids Res* **15**: 6799- 6811.

Ditta G, Stanfield S, Corbin D, Helinski DR (1980) Broad host range DNA cloning system for gram-negative bacteria: Construction of a gene bank of *Rhizobium meliloti*. *Proc Natl Acad Sci USA* **77**: 7347- 7351.

Fraenkel- Conrat (1983) RNA-dependent RNA polymerases in plants. *Proc Natl Acad Sci USA* **80**: 422- 424.

Fritig B, Rouster J, Kauffmann S, Stinzi A, Geoffroy P, Kopp M, Legrand M (1989) Virus-induced β-glycanhydrolases and effects of oligosaccharide signals on plant-virus interactions. In: Lugtenberg BJJ (ed) Molecular Signals in Microbe-Plant Symbiotic and Pathogenic Systems. Springer Verlag, Wien, in press.

Gianinazzi, S (1983) Genetic and molecular aspects of resistance induced by infection or chemicals. In: Nester EW and Kosuge T (eds) Plant-Microbe Interactions; Molecular and Genetic Perspectives, pp. 321- 342. Macmillan Publ. Co., New York.

Hilder VA, Gatehouse AMR, Sheermann SE, Barker RF, Boulter D (1987) A novel mechanism of insect resistance engineered into tobacco. *Nature* **300**: 160- 163.

Hoekema A, Hirsch PR, Hooykaas PJJ, Schilperoort RA (1983) A binary plant vector strategy based on separation of the Vir and T- region of the *Agrobacterium tumefaciens* Ti- plasmid. *Nature* **303**: 179- 180.

Hooft van Huijsduijnen RAM, Alblas SW, De Rijk RH, Bol JF (1986a) Induction by salicylic acid of pathogenesis-related proteins and resistance to alfalfa mosaic virus infection in various plant species. *J Gen Virol* **67**: 2135- 2143.

Hooft van Huijsduijnen RA , Kauffmann S, Brederode FTh, Cornelissen BJC, Legrand M, Fritig B, Bol JF (1987) Homology between chitinases that are induced by TMV infection of tobacco. *Plant Mol Biol* **9**: 411- 420.

Hooft van Huijsduijnen, RAM, Van Loon LC, Bol JF (1986b) cDNA cloning of six mRNAs induced by TMV infection of tobacco and a characterization of their translation products. *EMBO J* **5**: 2057-2061.

Horsch RB, Fry JE, Hoffman NL, Eicholtz D, Rogers SG, Fraley RT (1985) A simple and general method for transferring genes into plants. *Science* **227**: 1220-1231.

Hu N-T, Messing J (1982) The making of strand-specific M13 probes. Gene 17: 271-277.

Jamet E, Fritig B (1986) Purification and characterization of 8 of the pathogenesis-related proteins in tobacco leaves reacting hypersitively to tobacco mosaic virus. *Plant Mol Biol* **6**: 69-80.

Kauffmann S (1988) Les protéines PR (pathogenesis-related) du tabac: des protéines impliquées dans les réactions de défense aux agents phytopathogénes. Isolement, propriétés sérologiques et activités biologiques. PhD Thesis, Université Louis Pasteur, Strasbourg.

Kauffmann S, Legrand M, Geoffroy P, Fritig B (1987) Biological function of "pathogenesis-related" proteins. Four PR proteins of tobacco have 1, 3-ß-glucanase activity. *EMBO J* **6**: 3209- 3212.

Legrand M, Kauffmann S, Geoffroy P, Fritig B (1987) Biological function of "pathogenesis-related" proteins: four tobacco PR-proteins are chitinases. *Proc Natl Acad Sci USA* **84**:6750- 6754.

Linthorst HJM, Cornelissen BJC, Van Kan JAL, Van de Rhee M.D., Meuwissen RLJ, González Jaén MT, Bol JF (1989a) Induction of plant genes by compatible and incompatible virus-plant interactions. In: Fraser RSS (ed) Springer Verlag, Wien/New York, in press.

Linthorst HJM, Meuwissen RLJ, Kauffmann S, Bol JF (1989b) Constitutive expression of pathogenesis-related proteins PR-1, GRP and PR-S in tobacco has no effect on virus infection. *Plant Cell* **1**: 285-291.

Lotan T, Ori N, Fluhr R (1989) Pathogenesis-related proteins are developmentally regulated in tobacco flowers. *Plant Cell,* in press.

Mauch F, Mauch- Mani B, Boller T (1988) Antifungal hydrolases in pea tissue. II. Inhibition of fungal growth by combinations of chitinase and ß- 1, 3- glucanase. *Plant Physiol* **88**: 936- 942.

Memelink J, Linthorst HJM, Schilperoort RA, Hoge JHC (1989) Tobacco genes encoding acidic and basic pathogenesis-related proteins display different expression patterns. *Plant Mol Biol,* in press.

Patana R (1985) In: Sing P, Moore RF (ed) Handbook of Insect Rearing, Vol II, pp. 329-334 and pp 465- 468. Elsevier, Amsterdam.

Pierpoint WS (1986) The pathogenesis-related proteins of tobacco leaves. *Phytochem* **25**: 1595- 1601.

Pierpoint WS, Tatham AS, Pappin DJC (1987) Identification of the virus-induced protein of tobacco leaves that resembles the sweet protein thaumatin. *Physiol Mol Plant Pathol* **31**: 291- 298.

Richardson M, Valdes-Rodriguez S, Blanci-Labra A (1987) A possible function for thaumatin and a TMV-induced protein suggested by homology to a maize inhibitor. *Nature* **327**: 432- 434.

Shinshi H, Wenzler H, Neuhaus J- M, Felix G, Hofsteenge J, Meins F (1988) Evidence for N- and C-terminal processing of a plant defense-related enzyme: primary structure of tobacco prepro- ß- 1, 3-glucanase. *Proc Natl Acad Sci USA* **85**, 5541-5545.

Singh NK, Bracker CA, Hasegawa PM, Handa AK, Buckel S, Hermodson MA, Pfankoch E, Reguier FE, Bressan, RA (1987) Characterization of osmotin. A thaumatin- like protein associated with osmotic adaptation in plant cells. *Plant Physiol* **85**: 529- 536.

Smith CJS, Watson CF, Ray J, Bird CR, Morris PC, Schuch W, Grierson D (1988) Antisense RNA inhibition of polygalacturonase gene expression in transgenic tomatoes. *Nature* **334**: 724-726.

Van den Bulcke M, Bauw G, Castresana C, Van Montague M, Vandekerkhove, J (1989) Characterization of vacuolar and extracellular ß- 1, 3- glucanases of tobacco: evidence for a strictly compartmentalized plant defence system. *Proc Natl Acad Sci USA* **86**: 2673- 2677.

Van der Krol AR, Lenting PE, Veenstra J, Van der Meer IM, Koes RE, Gerats AGM, Mol JNM, Stuitje A (1988) An anti-sense chalcone synthase gene in transgenic plants inhibits flower pigmentation. *Nature* **333**: 866- 869.

Van Kan JAL, Cornelissen BJC, Bol JF (1988) A virus-inducible tobacco gene encoding a glycine-rich protein shares putative regulatory elements with the ribulose bisphosphate carboxylase small subunit gene. *Mol Plant-Microbe Interactions* **1**: 107-112.

Van Kan JAL, Van de Rhee D, Zuidema D, Cornelissen BJC, Bol JF (1989) Structure of tobacco genes encoding thaumatin-like proteins. *Plant Mol Biol* **12**: 153-155.

Van Loon LC (1985) Pathogenesis-related proteins. *Plant Mol Biol* 4: 111-116. Van Loon LC, Gerritsen YAM, Ritter CE (1987) Identification, purification and characterization of pathogenesis-related proteins from virus-infected Samsun NN tobacco leaves. *Plant Mol Biol* **9**: 593-609.

DISCUSSION OF J. BOL'S PRESENTATION

M. Zaitlin: You have shown that these PR proteins working alone apparently don't affect resistance, but are you are leaving the door open for them working in concert to bring about this effect?

J. Bol: Yes, of course, but the situation is rather complex. You have rather extensive gene families which are all in use, so in total many dozens of tobacco genes are in use following TMV infection. The fact that a single gene product is not effective does not rule out the possibility that you need a combination. It might be a hard job to make all possible combinations. For instance, the observation that salicylic acid treatments inhibit virus multiplication could give a clue, in that salicylic acid-inducible proteins would have a role in identifiable defense. Or because the PR-1 proteins and the GRP proteins are most strongly induced, we tested these for an antiviral response, but it could be possible that you need the combination of these two.

T. Hohn: Maybe you still have some effect (in insect resistance) and you could look at it by competition, so you give them two leaves, one being transgenic and the other not, maybe the insect prefers one of these. Did you do any experiments of that type?

J. Bol: No, we did not do that but I don't think that there is a very strong defense induced by TMV infection in tobacco against insect attack; because if you feed these larvae with TMV-inoculated leaves, the leaves are digested. We have done no competition experiments to see if they have any preference.

T. Hohn: I mean with your PR transgenic plants, that you compare those.

J. Bol: We have offered the larvae no choice between the two types of leaves.

R. Beachy: I'm concerned about the interpretation of the antisense data, because you have knocked down the amount of expression of the gene by 80 or 90%. From previous published data about what that means in the anti-polygalacturonase case, even having 10% remaining was enough to leave a lot of enzyme activity; thus, maybe the interpretation should be that 80 or 90% less doesn't affect it but maybe a 100% inhibition would do so?

J. Bol: Yes, of course, you are completely correct. It's just a suggestion that you can't find any function, but that doesn't rule out that there is a function of course. It is not known if these proteins have any enzymatic activity, but it could be possible.

R. Goldbach: Do you have any evidence for a trans-acting factor which could activate the promoters? Then instead of transforming a plant with a number of PR genes you would do better to clone one gene of this activator, which would then activate the normal genes of the plant itself. Do you have any evidence for that?

J. Bol: Well it is possible that an existing trans-acting factor is modified after infection. Usually reactions elicited by all types of elicitors, such as ethylene, respond very quickly in cell culture systems. You may have induction of the genes in a few minutes to one hour suggesting that the induction pathway is already in place, and perhaps it needs some modification of the trans-acting factor. But while our analysis of these trans-acting factors are still very preliminary it would be interesting to see if there is some competition between the various classes of genes to see if one factor is binding to all these TMV-inducible gene families, or if each family has its own set of trans-acting factors. There is still a long way to go, however.

Coat Protein Mediated Resistance in Transgenic Plants

Roger N. Beachy

Department of Biology, Washington University, St. Louis, MO 63130, USA

A variety of tools are available to the virologist for studies of the function of virus structural and non-structural proteins and for evaluating the effects of each protein on infection and disease. Perhaps the two most important are (1) the use of cloned cDNAs that, upon *in vitro* transcription, give rise to infectious transcripts and (2) the use of transgenic plants that express chimeric genes encoding virus sequences. By creating *in vitro* mutations in the virus cDNA it is possible to determine the role of each cistron in each stage of infection, replication, spread throughout the plant, and pathogenicity (virulence). Furthermore, transgenic plants can be used to examine the effects of individual virus proteins on resistance and susceptibility, and to rescue mutant virus (conditional or complete mutations) by *in trans* complementation. We are using both approaches to further characterize the genes of several viruses and determine their role(s) in resistance and susceptibility of tobacco (*Nicotiana tabacum*). In this report we summarize the results of experiments that involve expressing genes encoding the capsid proteins (CP) and the 30 kDa protein (movement protein, MP) of tobacco mosaic virus (TMV) in transgenic tobacco plants.

EXPRESSING VIRUS CAPSID PROTEINS IN TRANSGENIC PLANTS TO PRODUCE RESISTANCE

What is known about CP mediated resistance?

Although several reports in 1985 (Bevan *et al.*, 1985; Beachy *et al.*, 1985) described the expression of a TMV-CP gene in transgenic tobacco plants, it was not until 1986 that the impact of the CP on the plant was detailed. The report by Powell Abel *et al.*, (1986) demonstrated that transgenic tobacco plants that express a chimeric gene encoding the TMV-CP gene have increased resistance to infection by TMV. Since that report a number of other laboratories reported similar results when CP genes from other viruses were expressed in a variety of transgenic plants. The current list of virus CP genes, transgenic plants, and virus resistance is described in Table 1.

While the list of examples of CP-mediated resistance continues to grow, so does the understanding about how resistance is mediated. It is increasingly apparent that there are significant differences in the features of resistance in different host:virus systems. Some of these differences were described in a recent review by Hemenway *et al.*, (1989). We will confine this discussion primarily to the results with TMV-CP mediated resistance.

Table 1. Examples of coat protein mediated resistance in transgenic plants

CP Sequence*	Transgenic Plant	Virus Resistance	Reference
TMV	tobacco	TMV	Powell Abel *et al.*, 1986
	"	ToMV	Nelson *et al.*, 1988
"	"	PMMV	Nejidat & Beachy, unpublished
"	"	TMGMV	"
"	"	ORV	"
"	tomato	TMV	Nelson *et al.*, 1988
"	"	ToMV	"
ToMV	tomato	ToMV	Kaniewski *et al.*, unpublished
PVX	tobacco	PVX	Hemenway *et al.*, 1988
"	potato	PVX	Hemenway *et al.*, submitted
PVY	potato	PVY	"
SMV	tobacco	PVY	Stark & Beachy, 1989
"	"	TEV	"
AlMV	tobacco	AlMV	Loesch-Fries *et al.*, 1987;
"	"	"	Tumer *et al.*, 1987
"	"	"	van Dun *et al.*, 1987
"	tomato	"	Tumer *et al.*, 1987
CMV	tobacco	CMV	Cuozzo *et al.*, 1988
"	tomato	CMV	Cuozzo *et al.*, 1988
TRV	tobacco	TRV	van Dun and Bol, 1988
TSV	tobacco	TSV	van Dun *et al.*, 1987
"	"	PEBV	van Dun and Bol, 1988 *

*Abbreviations: TMV, tobacco mosaic virus; ToMV, tomato mosaic virus; PMMV, pepper mild mosaic virus; TMGMV, tobacco mild green mosaic virus; ORSV, ondontoglossum ringspot virus; PVX, potato virus X; PVY, potato virus Y; TEV, tobacco etch virus; AlMV, alfalfa mosaic virus; CMV, cucumber mosaic virus; TRV, tobacco rattle virus; TSV, tobacco streak virus; PEBV, pea early browning virus.

We have introduced chimeric genes encoding the CP gene from the U1 strain (common) of TMV into tobacco cv. Xanthi (a systemic host), cv. Xanthi *nc* (a local lesion host) and tomato (a systemic host) and studied various aspects of resistance. Powell *et al.*, (in press) carried out site-directed mutagenesis on the chimeric gene such that the transcript would include or not include the pseudoknot sequences at the 3' end of TMV-RNA, and would encode or not encode CP. A number of transgenic lines were produced and characterized to compare the levels of gene expression (mRNA levels) and/or CP accumulation and to assess the degree of resistance in these lines. An important conclusion from this work was that resistance was the result of CP accumulation. Furthermore, the degree of resistance to TMV was directly correlated with the amount of CP produced.

Resistance to infection is manifested in several ways. Nelson *et al.*, (1987) reported that there are fewer chlorotic or necrotic local lesions in plants that accumulate CP [i.e., CP(+)] than those that do not [i.e., CP(-)]. Register and Beachy (1988) reported that protoplasts from CP(+) plants were also resistant to infection. However CP(+) plants (Nelson *et al.*, 1987) and CP(+) protoplasts (Register *et al.*, 1988) were much less resistant to infection by TMV-RNA and by TMV that was "swollen" by treatment at pH 8.0 than to untreated TMV (Register *et al.*, 1988). These data lead to the conclusion that the CP(+) plants (and protoplasts) are less susceptible to infection than CP(-) plants and that resistance involves interference with an early stage of virus infection, perhaps the event(s) leading to the uncoating and release of viral RNA. Wilson *et al.*, (1989) demonstrated that the CP(+) protoplasts prevented the expression of a messenger RNA encoding β-glucuronidase if the mRNA was encapsidated by TMV-CP, a result that supports the conclusions of the experiments with TMV.

In addition to interfering with the establishment of infection, CP(+) plants permit less localized spread of viruses from the point of inoculation than do CP(-) plants (L.A. Wisniewski *et al.*, in preparation). Furthermore there is substantial reduction in the rate of systemic spread of virus in CP(+) plants compared with CP(-) plants (Powell Abel *et al.*, 1986; L.A. Wisniewski *et al.*, in preparation). This reduction in spread can in part be mimicked in plants that are grafted so as to contain a CP(+) stem section between CP(-) rootstock and apical sections. Finally, if systemic spread of virus takes place in CP(+) plants, the severity of the disease is considerably less than in CP(-) plants. In most cases mild disease symptoms are accompanied by reduced accumulation of virus. Taken together, the effects of the accumulation of CP in transgenic plants is to dramatically reduce infection and the severity of disease caused by TMV. It is unclear, however, whether the same or different mechanism(s) limit both infection and spread of TMV.

G. Clark *et al.*, (manuscript in preparation) gained additional insight into this question by using two different transcriptional promoters to drive expression of the CP gene. While the promoters from the 35S transcript of cauliflower mosaic virus (p35S) and a small subunit of ribulose bisphosphate carboxylase from petunia (pSSU) lead to equivalent levels of resistance in protoplasts, p35S:CP genes lead to considerably greater resistance to systemic spread than did pSSU:CP genes. This result indicates that the nature of the promoter, presumably the tissue and cell type specific expression of the CP gene can markedly affect the level of resistance in CP(+) plants.

A recurring question in TMV-CP mediated resistance relates to the nature of the molecule that confers protection. Results of experiments by Register and Beachy (in press) using a transient assay in protoplasts supports the conclusion that extended helical aggregates (i.e., rodlets) provide greater protection against infection than subunits or small aggregates of CP. Wilson and colleagues used extracts from the CP(+) plants described by Powell *et al.*, (1986) and demonstrated that they contain short rodlets of TMV-CP, some of which include encapsidated RNA (of unknown origin) (Wilson, 1989). The results of the experiments with protoplasts and the assays by Wilson imply that accumulation of higher order aggregates of the CP, for example rodlets, are involved in resistance.

The results pertaining to the role of rodlets of CP in protection led us to speculate that helical aggregates of the TMV-CP may provide protection not only against TMV but other tobamoviruses as well. Using the tobamovirus relatedness dendrogram described by Gibbs (1986) we undertook a study to compare resistance of the CP(+) tobacco plants against different viruses in this group. In this scheme of comparison tomato mosaic virus (ToMV) is most related to TMV, followed by pepper mild mottle virus (PMMV), tobacco mild green mosaic virus (TMGMV), ondontoglossum ringspot virus (ORSV), and then sunn hemp mosaic virus (SHMV). These studies showed that the TMV-CP gave very good protection against ToMV, PMMV and TMGMV, less resistance against ORSV, and considerably less resistance against SHMV (Nelson *et al.*, 1987; A. Nejidat and R.N. Beachy, in preparation). On the basis of these results we suggest that resistance against tobamovirus is due, at least in part, to structural similarities between the protecting CP (or macro-structure) and the challenge virus. Furthermore, the less related the virus, the less the resistance.

This hypothesis is further supported by research on CP-mediated resistance to a potyvirus by Stark and Beachy (1989). In this report tobacco plants were produced that express a chimeric gene encoding the CP of soybean mosaic virus (SMV), a potyvirus that does not infect tobacco. Progeny of the transgenic plant lines were challenged with tobacco etch virus (TEV) or potato virus Y (PVY), each of which is a pathogen of tobacco, and which have CP sequences that are approximately 60% homologous with the SMV CP. Plants that expressed the SMV CP gene were resistant to infection by each virus: one line was highly resistant even when the inoculum contained 50 or 100 μg virus/ml. Other lines were much less resistant, yet accumulated equivalent levels of SMV CP. While the reason for the different degrees of resistance are not understood, it is encouraging that a single potyvirus CP type can provide resistance against infection by several different potyviruses. Furthermore, this result supports the hypothesis (described above) that conferring protection against infection is dependent upon the degree of relatedness between the "protecting" CP and the challenge virus. Since the areas of high homology amongst CP sequences of the potyviruses lie in defined areas or regions (Shukla and Ward, 1989), it is likely that these areas are responsible for protein:protein interactions such as occurs in particle assembly. Such protein:protein interactions may be important in conferring protection.

What is unknown about CP-mediated protection?

Although new information about CP-mediated resistance is being gathered in a number of cases, much remains to be learned about the similarities and differences in protection against different viruses in different hosts. To date the systems most studied are the TMV-CP:tobacco and tomato systems. This is due in some measure to the extensive information about TMV structure, assembly and disassembly, and replication. In general, however, we lack knowledge about several important aspects of infection, spread, and disease development. Since the genetic phenotype of CP-mediated protection involves many stages, we find that the available information about disease limits our capacity to address the mechanism(s) of resistance. The positive aspect of this dilemma is that the transgenic CP(+) plants can help to answer questions about infection and disease as resistance is studied. Some of the remaining important questions include:

(1) How is virus infection initiated, and what are the primary steps in virus infection? Although the work of Wilson, Shaw and others have added substantially to our understanding of infection, additional experiments need to address how viruses disassemble. Concurrent with those studies it will be important to determine precisely how CP interferes with this process.

(2) Is there a CP-mediated effect on virus replication after virus infection is initiated? The experiments of Register and Beachy (1988) indicate that CP(+) protoplasts have a low but significant resistance against infection by TMV-RNA. Is this due to affects on replication *per se*, or to other factors? Additional experiments are needed.

(3) How does the accumulation of TMV (or other virus) CP limit the systemic spread of virus infection? Unfortunately, little is known about the precise nature of the spread of TMV from leaf mesophyll cells, which support most of the virus replication, through companion cells and into phloem, through which systemic virus movement occurs. Since the largest amount of CP synthesis in most of the transgenic CP(+) plant lines occurs in these cells (recall that the promoter used in nearly all of the transgenic plant studies to-date is the highly expressed cauliflower mosaic virus 35S promoter), it is likely that resistance to systemic spread is due to cell-specific gene expression. While this is supported by the results of experiments with the SSU promoter, additional experiments with cell-and tissue-specific transcriptional promoters will lead to better understanding of this resistance. Such studies will in return lead to greater understanding of virus spread in CP(-) plants.

THE ROLE OF THE TMV-MOVEMENT PROTEIN (30 K NONSTRUCTURAL PROTEIN) IN DISEASE

The second TMV protein receiving study in our laboratory is the non-structural 30 kDa protein. During the last 10 or so years several different experimental approaches lead to the hypothesis that the TMV 30 kDa protein is involved in some way with movement of TMV throughout the host. Two different experimental approaches, each involving the temperature-sensitive *LS1* strain of tomato mosaic virus, have documented that the 30 kDa protein is indeed responsible. Meshi *et al.*, (1987) demonstrated that a single amino acid change in the ORF representing the 30 kDa protein was responsible for the ts-nature of *LS1*. In a second approach Deom *et al.*, (1987) demonstrated that transgenic plants that express the 30 kDa protein from the common (U1) strain of TMV complement the *LS1* mutant virus. Based upon the role of this protein in virus spread it is referred to as the movement protein (MP) (based, in part, upon the suggestion by Palukaitis and Zaitlin, 1984).

Although the role of the MP in local and systemic spread of the virus has been thus described, little is known about how the MP functions. A paper by Tomenius *et al.*, (1987) clearly demonstrated that the MP is associated with plasmodesmata at the leading edge of the TMV infection. In an attempt to determine whether such an association can alter function of plasmodesmata, the transgenic plants that express the TMV-MP were used in microinjection experiments. In these studies intact leaves *in planta* were injected with fluorescent dyes of selected molecular mass, and the rate of spread of the dye was monitored. As a result of these studies Wolf *et al.*, (1989) found that plants that do not express the MP gene allow movement of fluorescent glucan molecules of ~750 Mr but not molecules of ~3900 Mr. In contrast, several MP(+) plant lines allow movement of molecules of 9400 Mr, but not 17,000 Mr. These results strongly indicate that the MP has a direct or indirect effect on plasmodesmata to alter size exclusion limits and, presumably, to enable virus or viral RNA to move from one cell to the next. However, much remains to be learned about the role of the MP in the changes that alter function of the plasmodesmata. It is, nevertheless, anticipated that the TMV-MP, and analagous proteins from other viruses, will provide new insight into cell- cell communication and non-infected as well as virus infected plants.

REFERENCES

Beachy, R. N., Abel, P., Oliver, M. J., De, B., Fraley, R. T., Rogers, S. G., and Horsch, R. B. (1985). Potential for applying genetic transformation to studies of viral pathogenesis and cross-protection. In: Biotechnology in Plant Sciences: Relevance to Agriculture in the 1980's. eds. M. Zaitlin, P. Day, and A. Hollaender, Academic Press, pp. 265-276.

Bevan, M.W., Mason, S.E., Goelet, P. (1985). Expression of TMV coat protein by a CaMV promoter in plants transformed by *Agrobacterium. EMBO J.* **4,** 1921-1926.

Cuozzo, M., O'Connell, K.M., Kaniewski, W., Fang, R.-X., Chua, N.-H., and Tumer, N.E. (1988). Viral protection in transgenic plants expressing the cucumber mosaic virus coat protein or its antisense RNA. *Bio/Technology* **6**, 549-557.

Deom, C. M., Oliver, M. J., and Beachy, R. N. (1987). The 30-kilodalton gene product of tobacco mosaic virus potentiates virus movement. *Science* **237**, 389-394.

Gibbs, A. (1986). Tobamovirus Classification. In "The Plant Viruses: Vol. 2 the Rod-shaped Plant Viruses." eds. M.H.V. van Regenmortel and H. Fraenkel-Conrat. Plenum Press, N.Y. pp. 168-180.

Hemenway, C., Fang, R.-X., Kaniewski, W.K., Chua, N.-H., and Tumer, N.E. (1988). Analysis of the mechanism of protection in transgenic plants expressing the potato virus X coat protein or its antisense RNA. *EMBO J.* **7**, 1273-1280.

Hemenway, C., Tumer, N. E., Powell, P. A., and Beachy, R. N. (1989). Genetic Engineering of Plants for Viral Disease Resistance. In Cell Culture and Somatic Cell Genetics of Plants. Vol. 6. Molecualr Biology of Plant Nuclear Genes. eds. J. Schell and I. K. Vasil. Academic Press., N.Y. pp. 406-424.

Loesch-Fries, L.S., Merlo, D., Zinnen, T., Burhop L., Hill, K., Krahn, K., Jarvis, N., Nelson, S., and Halk, E. (1987). Expression of alfalfa mosaic virus RNA4 in transgenic plants confers virus resistance. *EMBO J.* **6**, 1845-1851.

Meshi, T., Watanabe Y., Saito, T., Sugimato, A., Maeda, T., and Okada, Y. (1987). Function of the 30 kd protein of tobacco mosaic virus: Involvement in cell-to-cell movement and dispensability for replication. *EMBO J.* **6**, 2557-2563.

Nelson, R. S., Powell Abel, P., and Beachy, R. N. (1987). Lesions and virus accumulation in inoculated transgenic tobacco plants expressing the coat protein gene of tobacco mosaic virus. *Virology* **158**, 126-132.

Nelson, R. S., McCormick, S. M., Delannay, X., Dubé P., Layton, J., Anderson, E. J., Kaniewska, M., Proksch, R. K., Horsch, R. B., Rogers, S. G., Fraley R. T., and Beachy, R. N. (1988). Virus tolerance, plant growth, and field performance of transgenic tomato plants expressing coat protein from tobacco mosaic virus. *Bio/Technology* **6**, 403-409.

Palukaitis, P., and Zaitlin, M. (1984). A model to explain the "cross- protection" phenomenon shown by plant viruses and viroids. In "Plant- Microbe Interaction: Molecular and genetic perspectives" (T. Kosuge and E.W. Nesteer, eds.), Chap. **17**, pp. 420-429. Macmillan, New York.

Powell Abel, P.A., Nelson, R. S., De, B., Hoffmann, N., Rogers, S. G., Fraley R. T., and Beachy, R. N. (1986). Delay of disease development in transgenic plants that express the tobacco mosaic virus coat protein gene. *Science* **232**, 738-743.

Powell, P. A., Sanders, P. R., Tumer N., and Beachy, R. N. (1990). Protection against tobacco mosaic virus infection in transgenic plants requires accumulation of capsid protein rather than coat protein RNA sequences. *Virology* (in press).

Register, J. C. III, and Beachy, R. N. (1988). Resistance to TMV in transgenic plants results from interference with an early event in infection. *Virology* **166**, 524-532.

Register, J. C. III, and Beachy, R. N. (1989). Effect of protein aggregation state on coat protein-mediated protection against tobacco mosaic virus using a transient protoplast assay. *Virology* **173**, 656-663.

Shukla, D. D., and Ward, C. W. (1989). Structure of potyvirus coat proteins and its application in the taxonomy of the potyvirus group. *Adv. in Virus Res.* **36**, 273-314.

Stark, D. M., and Beachy, R. N. (1989). Protection against potyvirus infection in transgenic plants: Evidence for broad spectrum resistance. *Bio/Technology* (in press).

Tomenius, K. D., Claphan, D., and Meshi, T. (1987). Localization by immunogold cytochemistry of the virus-coded 30K protein in plasmodesmata virus. *Virology* **160**, 363-371.

Tumer, N.E., O'Connell, K.M., Nelson, R.S., Sanders, P.R., Beachy, R.N. (1987). Expression of alfalfa mosaic virus coat protein gene confers cross-protection in transgenic tobacco and tomato plants. *EMBO J.* **6**, 1181-1188.

van Dun, C.M.P., Bol, J.F., and van Vloten-Doting, L. (1987). Expression of alfalfa mosaic virus and tobacco rattle virus coat protein genes in transgenic tobacco plants. *Virology* **159**, 299-305.

van Dun C.M.P., and Bol, J. F. (1988). Transgenic tobacco plants accumulating tobacco rattle virus coat protein resist infection with tobacco rattle virus and pea early browning virus. *Virology* **167**, 649-652.

Wilson, T. M. A. (1989). Plant viruses: A tool-box for genetic engineering and crop protection. *Bio Essays* **10**, 179-186.

Wolf, S., Deom, C. M., Beachy, R. N., and Lucas, W. J. (1989). Movement protein of tobacco mosaic virus modified plasmodesmatal size exclusion limit. *Science* **246,** 377-379.

DISCUSSION OF R. BEACHY'S PRESENTATION

B. Harrison: In relation to your PVY potyvirus coat protein transformants, have you tried experiments by a graft challenge of those plants, and secondly, have you tried to put on the top of the transgenic plant a graft of a infected non-transgenic plant to see what happens to the virus content in that tissue?

R. Beachy: No we have not tried to do graft transmission of virus in these plants, either with TMV or with the potyviruses. The only experiments looking at alternative methods of transmission are those done by Wodciech Kaniewski who has done the transmission by aphids and finds resistance against aphid transmission of PVY, but we have not done the graft transmission of virus in those cases. It's a good experiment.

R.Hull: Do you think that a gateing capacity of less than 17,000 would be big enough to let an infectious unit of TMV through?

R. Beachy: The biggest question of course is what is the infectious unit. If the infectious unit is RNA which can move between cells, then perhaps it could be globular enough to snake through. You think of a plasmodesmata as a channel about 50 nm wide which has in it a lot of proteins which together reduce the effective size of that to about 0.1 nm. What we are asking is how much of a change needs to happen to make it large enough for viral RNA to move through, then the suggestion would be we would have to get to the range of 5-10 nm to let a globular RNA molecule move. I don't know if what we have demonstrated shows that much of a change and it certainly wouldn't be enough of a change to allow a virus particle to move. Remember that the assays are of a neutral molecule, not a charged one, and we have no idea what might happen if you use not a neutral dextran but another charge as well. All we are looking at is one component. It is possible that this is enough to do it, and certainly the work that Curtis Holt has done with his infectious clone would indicate that it certainly complements that mutation and allows it to move without having more of it made.

M. Zaitlin: Isn't the hydrated volume of those dextran adducts rather large? It's in the nm class isn't it?

R. Beachy: The estimate for the hydrated size of the dextrans varies between 1 and 10 nm, or 1 and 7 nm at least. So it's unknown what it is in the state that we have it in the cell, and it may be somewhat different than what we have on a Sephadex column. We simply don't know. I think the telling thing will be to now inject lots of things in this system. Of course you can inject double-stranded RNAs or DNAs or single-stranded RNAs. You can do the experiments now; the experimental design is set up so that you can inject whatever you like as long as you can attach it to a fluorescent molecule. So you can see we can do lots of things in the injection process, adding in molecules that might block the function of plasmodesmata, or might accelerate it as Professor Atabekov suggests; perhaps cAMP or the inhibitors of cAMP would be ones that should be injected. All of the system is now available to not only manipulate the virology but the cell biology of plasmodesmata and the whole thing is wide open as to what might be able to move now that these channels are altered enough to allow nucleic acid or viral infectivity to move.

W. Gerlach: Have you ever seen any protection using truncated forms of coat proteins for any of these viruses? Or say with TMV ones which do not form rodlets?

R. Beachy: The experiments that involved truncation of the proteins or making point mutations have unfortunately not gone quite as well as they should have. So we have no data on those forms. Dave Stark had made some deletions of the potyviruses. As you might expect he deleted the part of the protein that sticks out at the end that is trypsin sensitive. That has been deleted, as have been two other deletions that go into those highly conserved cores. We have transgenic plants that are coming back from those. Some of those expressed the truncated protein and can now be assayed for protection, but it will be another six months before we have the complete story on those plants and their degree of resistance.

M. Wilson: You were alluding to these constructions with the small subunit promoter and the 35S promoter with the coat protein? Do you have any gold-labeling data, as you speculated, localizing the coat protein on different sites?

R. Beachy: I didn't say different sites, I said different tissue types. I was careful not to say different sites because we don't have those data yet. It needs to be done very carefully to see if there are differences in subcellular localizations as well. Not simply in tissue. In fact, if you do the tissue printing experiments with these plants that have the 35S promoter driving the coat protein gene, you can see that it is primarily in phloem and phloem-associated cells. You can pick that up. But in the case of the small subunit there is not even enough there to pick up, presumably because it is so uniformly spread. But the electron micrographs do need to be done. Another reason is that when we did these coat protein transformants with the soybean mosaic virus coat protein gene, Dave Stark looked at a lot of lines, many of which have the same level of expression and protein accumulation but the level of resistance was markedly different. We have several lines that are nearly immune, other lines that are at a very intermediate level of resistance and other lines that express constant protein that have almost no protection. So I don't know what that means but the resistance is virus group specific. In other words you challenge those plants with TMV and they are as susceptible as the normal. So it is a virus specificity in the resistance, but we certainly don't understand all the features in the potyvirus case that lead to resistance, and I think if you talk to Nilgun Tumer and Wodciech Kaniewski I think you will find similar things in the potato case. It is a very interesting puzzle and I think to understand it it's going to take a lot more biochemistry and cell biology.

M. Zaitlin: Over the years there have been all kinds of studies about viruses in plasmodesmata and there have been reports that they perturb the plasmodesmata, that desmotubules are perturbed, and there is some suggestion that plasmodesmata come and go. Have you looked at any of this with respect to these transgenics in the upper and lower leaves?

R. Beachy: We have not looked at electron micrographs in the upper and lower leaves. The only thing we have done is to look at tissue prints and the rest of the material is just simply not finished yet. It is a good experiment because there are examples in the literature of finding TMV in plasmodesmata or presumably in plasmodesmata. There is the whole question of whether or not these are the same plasmodesmata, i.e., whether you are modifying existing plasmodesmata, or making new ones or newly modified structures which together with the virus infection cause something quite different than the normal plasmodesmata. We don't know what transmits virus. We only have this assay that says we do have some change in plasmodesmatal function. Whether or not it changes enough in the severe case or perhaps in a very aged infection in a leaf that is not very happy and is beginning to senesce isn't known; maybe in those examples you would have virus in the channels which remain open. I would be surprised if it would be a function of 90% of the plasmodesmata to have virus particles in them. Perhaps a very small percentage would not be unexpected. And remember there are hundreds of plasmodesmata around a single cell, not just a few. So there are lots of ways we can think a small percentage might be modified, but not all of them. Or they may become all modified as the plant ages. So there are a lot more questions to answer than there might seem to be at first.

Structure and Functions of Tobacco Mosaic Virus RNA

Yoshimi Okada*, Tetsuo Meshi, and Yuichiro Watanabe

Department of Biophysics and Biochemistry, Faculty of Science, University of Tokyo, Hongo, Tokyo, Japan

Tobacco mosaic virus (TMV) is one of the well-characterized plant viruses. The genome of TMV is a positive-sense, single-stranded RNA and encodes at least three non-structural proteins (130K, 180K and 30K proteins) and a coat protein (Goelet *et al.*, 1982; Ohno *et al.*, 1984). The functions of non-structural proteins are not well understood at the molecular level. Recently, *in vitro* expression systems that allow production of infectious TMV RNAs from cloned full-length cDNA copies have been established (Dawson *et al.*, 1986; Meshi *et al.*, 1986) and as a result reverse genetics approaches have become possible for TMV research. We have constructed several kinds of TMV mutants *in vitro* to identify the function of TMV-coded proteins and the genomic RNA.

The 130K and 180K proteins were suggested to be involved in viral RNA replication (Haseloff *et al.*, 1984, Hunter *et al.*, 1976, Kamer and Argos, 1984). We introduced several kinds of mutations into TMV RNA at or near the amber termination codon of 130K protein gene (Ishikawa *et al.*, 1986). Mutants that did not produce the 180K protein were not infectious. In the case of mutants producing the 180K protein but an undetectable amount of the 130K protein, infectivity was retained, although its specific activity was very low. Balanced expression of the 130K and 180K proteins seems necessary for efficient replication of TMV RNA.

We constructed various kinds of TMV mutants of which the 30K protein genes were modified or deleted, and analyzed their biological properties in both tobacco plants and protoplasts (Meshi *et al.*, 1987). Analyses of these mutants have confirmed the previous suggestion (Leonard *et al.*, 1982) that the 30K protein is involved in cell-to-cell movement of virus. The same conclusion has been presented using transgenic tobaccos (Deom *et al.*, 1987). The 30K protein, as well as the coat protein, is dispensable for replication of TMV RNA (Meshi *et al.*, 1987). Comparisons among the 30K proteins of tobamoviruses show that the 30K proteins contain a rather conserved N-terminal two-thirds and a less-conserved C-terminal one-third (Saito *et al.*, 1988). The 30K protein has been localized in plasmodesmata of infected tobacco leaves by immunogold cytochemistry (Tomenius *et al.*, 1987). The 30K protein seems to accumulate in

* Present address: Department of Biosciences, School of Science and Engineering, Teikyo University, Utsunomiya, Tochigi, Japan

plasmodesmata before significant amounts of viral antigen begin to pass through them. The 30K protein may function early in cell-to-cell movement of virus.

When a coatless TMV mutant was inoculated on to a systemic host, no TMV related RNAs or proteins could be detected in the upper uninoculated leaves one week post-inoculation (Takamatsu *et al.*, 1987). This supports the idea (Siegel *et al.*, 1962) that the coat protein is involved in long-distance movement of virus. The coat protein gene encodes a factor responsible for the induction of the hypersensitive response in the N' plants (Saito *et al.*, 1987; Knorr *et al.*, 1987; Culver *et al.*, 1989). Some mutants in the coat protein gene caused necrotic local lesions or chlorotic spots on the inoculated leaves of the nn' tobacco plants (Dawson *et al.*, 1988; Saito *et al.*, 1989). The coat protein of TMV appears to have a potential in nature to induce several symptoms in infected leaves.

The 3' terminal portion of the genomic RNA is thought to be important for the specific initiation of viral minus-strand RNA synthesis during replication of viral RNA. To understand the specific interaction between TMV replicase and the 3' non-coding sequence, three TMV-L (tomato strain)-derived chimeras were constructed by replacing the 3' non-coding region with the corresponding sequence of TMV-OM (common strain), TMV-Cc (cowpea strain), or cucumber green mottle mosaic virus. All the chimeric viruses constructed were replication-competent, which indicates that some looseness is permitted in the interaction between TMV replicase and the 3' terminal region (Ishikawa *et al.*, 1988).

TMV resistance genes, *Tm-1* and *Tm-2* in tomato, have been identified and used in practice to prevent systemic mosaic symptoms by TMV (Fraser *et al.*, 1985). In tomato plants with these genes, multiplication of TMV is inhibited. To understand the resistance mechanisms, characterization of resistance-breaking strains (virulent) and subsequent comparisons with avirulent, parent strains were performed. The results showed strong correlations between the ability to overcome the *Tm-1* resistance and a few amino acid substitutions of the 130K and 180K proteins (Meshi *et al.*, 1988) and between the ability to overcome the Tm-2 resistance and alteration of the 30K protein (Meshi *et al.*, 1989).

In this symposium, our latest results on the structure and functions of TMV RNA will be presented, although most are still in progress.

30K PROTEIN

The C-terminal region of the 30K protein is not necessary for the function of cell-to-cell movement

Alignment of the 30K protein sequences of several tobamoviruses reveals two relatively well-conserved regions in the middle portion of the 30K protein, which are designated regions I and II and are probably important for the cell-to-cell movement function (Saito *et al.*, 1988). Temperature-sensitive mutations for cell-to-cell movement are found in one of the two well-conserved regions (Meshi *et al.*, 1987; Ohno *et al.*, 1983; Zimmern and Hunter, 1983). Amino acid changes responsible for overcoming the *Tm-2* resistance have been mapped also in or close to the well-conserved regions (Meshi *et al.*, 1989).

The C-terminal less-conserved region can be divided into three subregions based on the distribution of charged residues: regions A, B and C. Regions A and C are rich in acidic amino acids, while region B is rich in basic amino acids (Saito *et al.*, 1988). Although the C-terminal one-third of the 30K protein is less conserved, distribution of charged amino acid residues in this region is similar among tobamoviruses.

Proteins homologous to the 30K protein have been reported from viruses other than tobamoviruses. Homology between the 30K protein of the TMV and the 29K protein of tobacco rattle virus (TRV), a member of the tobravirus group, has been shown (Boccara *et al.*, 1986; Cornelissen *et al.*, 1986). Similarities are detected through the entire sequence, and regions I and II are relatively conserved. This may indicate that tobamoviruses and tobraviruses share a common transport mechanism. The C-terminal region of the 29K protein of TMV is shorter than that of the 30K protein of TMV and lacks region C and the latter half of region B. To examine whether the entire sequence of the 30K protein of TMV is necessary for the cell-to-cell movement function, we constructed a series of mutants with deletions in the C-terminal coding region of the 30K protein.

The 30K protein of TMV-L is composed of 264 amino acid residues. A mutant, designated as D233, has the coding sequence for the N-terminal 233 amino acid residues but not for the C-terminal 31 amino acid residues which contain region C and the latter half of region B (Saito, T., Meshi, T., and Okada, Y., unpublished result). D233 produced necrotic local lesions on Xanthi nc tobacco leaves, indicating that at least the C-terminal 31 amino acid residues of the 30K protein of TMV are dispensable for the cell-to-cell movement function. Experiments on other mutants having longer C-terminal deletion are now in progress.

Localization of the 30K protein in tobacco protoplasts infected with TMV RNA

Previously, the 30K protein of TMV has been localized in plasmodesmata of TMV-infected tobacco leaves by immunogold cytochemistry (Tomenius *et al.*, 1987). In the experiment, affinity-purified antibodies against a synthetic hexadecapeptide corresponding to the C-terminal sequence of the 30K protein were used (Ooshika *et al.*, 1984). Thus the sites of the 30K protein immunoreactive with these antibodies must be restricted to the C-terminal region corresponding to the synthetic peptide, the region dispensable for cell-to-cell movement. In addition, antibodies would not react with the 30K protein of which the C-terminal region might be buried in certain structures or which might not remain intact. Therefore we prepared antibodies that were expected to react with various portions of the 30K protein molecule.

For this, we chose to use 30K protein expressed in *E. coli* as an immunogen. We cloned cDNA derived from the open reading frame of the 30K protein gene into the pET-3b vector, which had been developed for expression of target DNAs under control of the T7 promoter (Rosenberg *et al.*, 1987), and constructed pLQV4 and pLQV54. pLQV5 was designed to produce the same protein as the authentic 30K protein, and pLQV4 a 30K protein derivative with 11 extra amino acids attached to the N-terminus. *E. coli* BL21 (DE3) lysogens containing pLQV4 or pLQV5 were cultured essentially according to the method of Studier *et al.*, (1989). The 30K proteins synthesized in *E. coli* were easily isolated as aggregates by sonication and subsequent brief centrifugation. Proteins derived from pLQV4 and pLQV5 were designated ep31 and ep30, respectively. These *E. coli*-expressed proteins were solubilized in the presence of SDS and 2-mercaptoethanol and separated from other proteins by SDS-PAGE.

Rabbits were immunized with ep31 which had been eluted from SDS-polyacrylamide gels. The IgG fraction was further purified by means of ep30-adsorbed nitrocellulose filters. Reactivity and specificity of antibodies were first examined using *in vitro* translation products derived from the *in vitro* transcripts carrying the 30K protein gene. The antibodies specifically immunoprecipitated the 30K protein. Reactivity with the 30k protein synthesized *in vivo* was then confirmed.

Protoplasts prepared from a tobacco suspension culture were inoculated with TMV RNA by electroportion (Watanabe *et al.*, 1987). The 30K protein is transiently synthesized at an early stage of infection in TMV-infected protoplasts (Watanabe *et al.*, 1984). Protoplasts were fixed at this early stage of infection. Sections were made and treated with the purified anti 30K protein (ep31) antiserum followed by the protein A-gold complex. Gold label was dispersed over the cytoplasm at this stage as shown in Fig. 1 (Hosokawa, D., Kawagishi, M., Meshi, T., Watanabe, Y., and Okada, Y., unpublished data). Time course analyses of localization of the 30K protein in protoplasts at later stages are now in progress.

The 30K protein has previously been localized in plasmodesmata of TMV-infected tobacco leaves (Tomenius *et al.*, 1987). On the other hand, the 30K protein was reported to be found in a nuclei-rich or fast-sedimenting fraction of TMV-infected protoplasts after subcellular fractionation of protoplast lysates (Watanabe *et al.*, 1986). This immunogold cytochemical method showed that the 30K protein may stay in the cytoplasm and might be associated with some cellular structure after it is synthesized, since protoplasts lack both cell walls and plasmodesmata.

Recently we have succeeded in regeneration of TMV-infected tobacco mesophyll protoplasts into tobacco plants. Further studies on the localization of the 30K protein during the course of regeneration from protoplasts may give

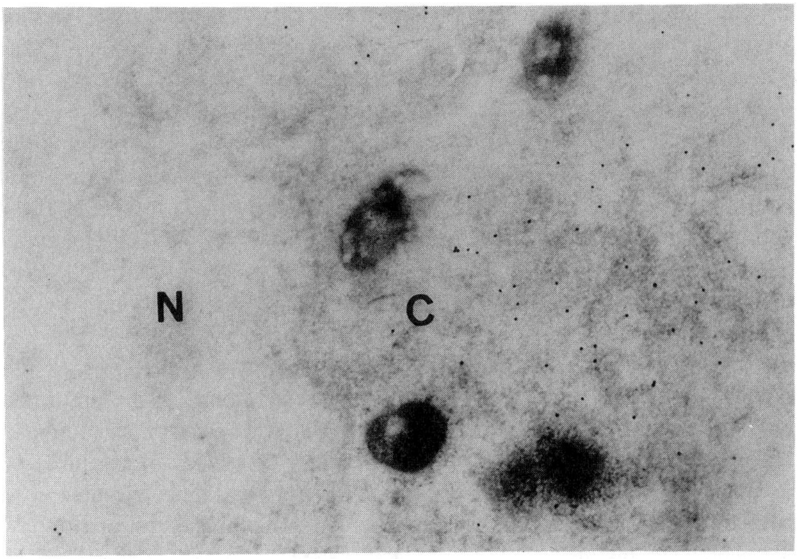

Fig. 1 Immunocytochemical labeling of the 30K protein in TMV-infected tobacco protoplasts. N, nucleus; C, cytoplasm.

some important information on the transport of the 30K protein from cytoplasm to plasmodesmata.

COAT PROTEIN

Hypersensitive response in tobacco plants with the N'gene

A tomato strain (L) of TMV induces a hypersensitive response in tobacco plants with the N' gene. A factor responsible for the induction of the hypersensitive response has been mapped in the coat protein gene (Saito *et al.*, 1987). We have constructed several mutants which have insertions or deletions in the coat protein gene (Saito *et al.*, 1989). Frame-shift mutants resulting in premature terminations of translation of the coat protein caused no necrotic local lesions of N' plants. Mutants which resulted in the expression of coat protein derivatives with one amino acid inserted after residue 101 or 152 caused necrotic local lesions on N' plants. Deletion mutants lacking fewer than the C-terminal 13 amino acid residues in the coding region caused necrotic local lesions, while mutants lacking in the coding region for the C-terminal 38 amino acid residues caused no necrotic local lesions.

The coat proteins of TMV-L and TMV-OM, Japanese tomato and common strains respectively, are both composed of 158 amino acid residues, and there are 27 amino acid changes between them (Takamatsu *et al.*, 1983). To know which region of the coat protein gene of TMV-L is responsible for the induction of the hypersensitive response, we constructed two chimeras of the coat protein gene, in which the regions coding for the N-terminal two-thirds and the C-terminal one-third of the protein were separately derived from TMV-Om and TMV-L (Saito *et al.*, 1989). Both TMV chimeras caused necrotic local lesions on N' plants. This indicates that the sites responsible for induction exist in both coding regions for the N-terminal two-third and for the C-terminal one-third of the coat protein in the case of TMV-L. These results suggest that an entire structural feature rather than a small part of the coat protein might be important for the induction of the hypersensitive response in the N' plants.

Involvement in long-distance movement

Involvement of the coat protein in long-distance movement has long been presumed (Siegel *et al.*, 1962). We have shown that a coatless mutant lacking most of the coat protein gene is defective in long-distance movement (Takamatsu *et al.*, 1987). To investigate whether the ability to assemble into the virus particle is correlated with that of long-distance movement, we constructed various kinds of mutants (Saito *et al.*, unpublished observations).

A mutant of which the coat protein lacked the C-terminal 5 amino acids and mutants whose coat proteins had one amino acid insertion after residue 101 or 152 retained the capacity for long-distance movement and the ability to assemble into virus particles. Other mutants that expressed shorter coat protein derivatives or derivatives with one amino acid insertion after residue 56 lost both features. These results suggest that coat protein with the ability to assemble into virus particles may be required for long-distance movement.

To know more clearly the correlation between the ability to form particles and long-distance movement, we planned to construct a mutant that would produce the normal coat protein but could not form virus particles, by introducing multiple point mutations into the assembly origin. The assembly origin of TMV-L, which is in the 30K protein gene, can be folded into a highly base-paired hairpin-loop structure (Takamatsu *et al.*, 1983). As shown in Fig. 2, mutations were then selected so that the secondary structure would be maximally changed without alterations of the amino acid sequence of the 30K protein. The mutant RNA did not detectably assemble with the coat protein under *in vitro* assembly conditions. This mutant is referred to as FAD. When the FAD RNA was inoculated onto Xanthi nc leaves, the necrotic local lesions caused were nearly identical to those caused by the parental TMV-L. In the upper, uninoculated leaves of Samsun tobacco, a systemic host, this mutant did not cause any clear mosaic symptoms, though it occasionally caused a few chlorotic spots (Fig. 3).

Propagation and long-distance movement of FAD were examined by determining the accumulation of viral products. In the inoculated leaves, levels of accumulation of the coat protein were similar between FAD and TMV-L. However, the genomic RNA of FAD accumulated to about one-tenth the extent of that of TMV-L and this reflects the assembly-defective nature of the FAD RNA. In the upper leaves, the accumulation of the genomic RNA of FAD was about 100-fold less than that of TMV-L. Accumulation of the coat protein in the upper leaves was also reduced (about ten-fold less than that of TMV-L), and it was detected only in the areas including chlorotic spots. The number of infectious entities that moved into the upper leaves from the inoculated leaves might be much smaller in plants inoculated with FAD than in plants inoculated with the wild-type TMV. The result indicated that mutations in the assembly origin affect the efficiency of viral long-distance movement. Therefore, the sequence of the assembly origin on the genomic RNA is involved in long-distance movement as a *cis*-acting element.

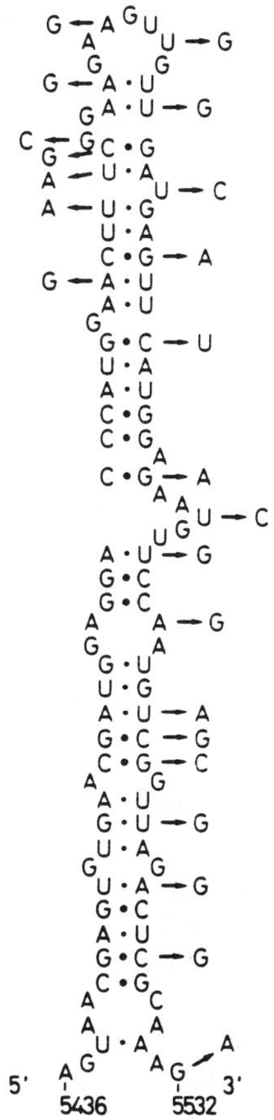

Fig. 2 Nucleotide sequence of the assembly origin of TMV-L and introduced mutations. A hairpin structure of the assembly origin formed by nucleotide residues 5436-5532 from the 5' -terminus of the genomic RNA are shown. Nucleotide residues indicated by arrows are mutated. Dots are putative base-pairing sites. (From Saito et al., 1990, by permission).

Fig. 3 Symptoms caused by an assembly origin-modified mutant FAD. Upper panel shows necrotic local lesions on leaves of Xanthi nc tobacco inoculated with FAD or W3 (TMV-L). Lower panel shows chlorotic spots (indicated by arrows) on the upper leaves of Samsun tobacco infected by FAD.

The *in vivo* viral assembly of FAD was analyzed. UV-monitoring and electrophoretic analysis after sucrose density gradient centrifugation could not demonstrate the existence of virus particles in extracts of inoculated leaves (Okada *et al.*, 1988), but fractions corresponding to the peak fractions containing standard TMV were shown by infectivity assay and electron microscopic observation to contain a very small amount of virus particles. Based on infectivity, the amount of virus particles contained in these fractions was deduced to be about one-five hundredth that of TMV-L in the corresponding control fractions. Inefficient assembly of virus particles would explain the inefficient long-distance movement of FAD. The form in which the virus moves may be the virus particles, although the possibility cannot be excluded that another complex, such as an informosome-like ribonucleoprotein complex (Dorokhov *et al.*, 1984), may be the form for long-distance movement. In the latter case, not only the coat protein but the assembly origin sequence must be involved in the formation of the ribonucleoprotein complex.

NON-CODING REGIONS

3' Non-coding region

Of the 202-nucleotide-long 3' non-coding region of TMV-L, the 3'-terminal portion of the 105 nucleotides (residues 6280-6384) can be folded into a tRNA-like structure (Rietvelt *et al.*, 1984) and the immediately upstream sequence of 75 nucleotides (residues 6203-6277) contains three pseudoknot structures (Van Belkum *et al.*, 1985). Such structures can be found in all the tobamovirus RNAs of which the 3'-terminal sequences have been determined. In the case of brome mosaic virus (BMV), the tRNA-like structure has been shown to play an essential role in replication of viral RNA (Dreher and Hall, 1988; Dreher *et al.*, 1989; Bujarski *et al.*, 1986), and it is supposed that this can apply to TMV. To know the biological function of the well-conserved, highly-structured pseudoknot region, we constructed mutant TMV RNAs with various lengths of nucleotide sequence of deleted from the pseudoknot region, and analyzed their multiplication in both tobacco plants and protoplasts (Takamatsu, N, Watanabe, Y., Meshi, T., and Okada, Y., unpublished results).

The three pseudoknots are, in order, referred to as the 5' (residues 6203-6225), central (residues 6226-6247) and 3' (residues 6248-6277) pseudoknots. Serial deletions were introduced just downstream of the termination codon of the coat protein gene in the 5'-to-3' direction progressively (Fig. 4). The 5'-proximal and central pseudoknot regions were dispensable for virus multiplication. However, extension of the deletion into the central pseudoknot region resulted in substantial reduction of multiplication, accompanied by loss of mosaic symptom development on systemic tobacco plants. For example, the N-6209 mutant RNA from which residues 6188-6209 were deleted showed typical systemic mosaic symptoms in the upper, uninoculated leaves about one week post-inoculation as did the parental TMV RNA. As for the N-6233 mutant (residues 6188-6233 deleted), systemic mosaic symptoms did not develop even one month post-inoculation, but infectivity, which was one-fifth or less than that of the wild type TMV, was recovered from the upper, uninoculated leaves, indicating the systemic spread of the mutant.

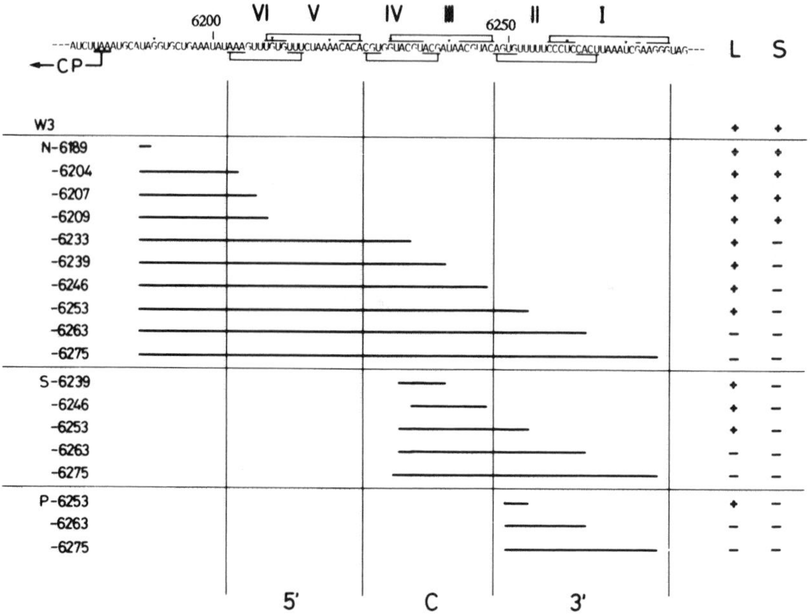

Fig. 4 The sequence of the pseudoknot region of TMV-L RNA and a schematic representation of individual deletions. The sequence is numbered from the 5' end. Residues underlined or overlined are those involved in base-pairing. The 5', central (C) and 3' pseudoknots which are indicated at the bottom consist of the double helical segment VI and V, IV and III, or II or I, respectively. The extent of deleted sequences for each mutant is represented by a solid line. At the right are shown the results of inoculation assays using the local lesion (L) and systemic (S) hosts. + indicates production of local lesions (in column L) or development of systemic mosaic symptoms at two weeks post-inoculation (in column S).

The deletions down to residue 6263, which is at the center of the 3'-proximal pseudoknot region, abolished viability. To determine whether the loss of viability for the N-6263 mutant RNA was due to lack of some sequence necessary for multiplication or simply due to the large extent of deletion, we constructed the S- and P-series mutant RNAs whose deletion started at the middle of the central pseudoknot and the 5' part of the 3' -proximal pseudoknot region, respectively (Fig. 4). In both series of mutant RNAs, the deletions which included residue 6263 resulted in abortion of viral viability irrespective of the extent of deletion. These results show that at least the 3' half of the 3'-proximal pseudoknot region if necessary for multiplication.

5' Non-coding region

The 5' untranslated leader sequence of TMV RNA forms an initiation complex with 80S ribosomes to translate the 130K and 180K proteins. In addition to the initiation codon (AUG), an AUU within this region, 51 bases upstream of the AUG codon, has been implicated in the binding of an 80S ribosome (Tyc *et al.*, 1984). Further, the 5' untranslated leader sequence of TMV has been shown to be a general enhancer of eukaryotic and prokaryotic translation *in vitro* (Gallie *et al.*, 1987a) and *in vivo* (Gallie *et al.*, 1987b). Mutational analysis has been carried out to determine those sequences necessary for translational enhancement (Gallie *et al.*, 1988). The complementary sequence of the 5'-noncoding region is also supposed to play an important role in initiation of plus-strand synthesis during viral RNA replication. To understand the biological function of this region, we constructed mutant RNAs with various lengths of deletions and analyzed their multiplication in protoplasts and also the expression of the 130K protein *in vitro* (Takamatsu, N., Watanabe, Y., Iwasaki, T., Meshi, T., and Okada, Y., unpublished results).

Unexpectedly, the sequence of residues 9-47 containing the AUU (Tyc *et al.*, 1984) was dispensable for viral multiplication in tobacco protoplasts (Fig. 5). However, deletion of residues 26-72 abolished viability. These mutant RNAs were translated in rabbit reticulocyte lysate, and the 130K protein was detected by SDS-polyacrylamide gel electrophoresis even in the case of non-viable mutant RNA. These results suggest that the loss of viability of TMV RNA having a deletion of residues 26-72 (this mutant is referred to as [25/73]

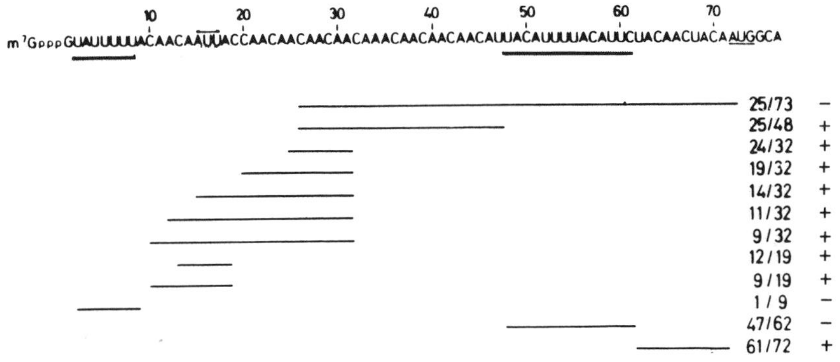

Fig. 5 The sequence of the 5' untranslated leader sequence of TMV-L RNA and a schematic representation of individual deletions. The sequence is numbered from the 5' end. AUG. underlined is the initiation codon of the 130K protein. Sequences boldly underlined are the sequences indispensable for viral viability. The extent of deleted sequences for each mutant is represented by a solid line. At the right are shown the results of inoculation assays using tobacco protoplasts. + indicates replication of the inoculated RNA.

was not due to the loss of an ability to translate the 130K protein gene. To determine whether the inviability of TMV RNA [25/73] was due to the lack of some essential sequence for multiplication or simply due to the deletion of a long sequence, we constructed two mutant RNAs, [47/62] lacking the sequence of residues 48-61 and [61/72] lacking residues 62-71. The mutant RNA [47/62] was not viable, while RNA [62/72] multiplied in protoplasts. This shows that the sequence of residues 48-61 contains that responsible for viability.

At the same time, we constructed a mutant RNA which had a deletion in the 5'-proximal region of residues 2-8. This mutant [1/9] was inviable as expected. These results show that the sequences necessary for multiplication are contained in the two separate residues 1-8 and 48-61. Since the sequence of residues 1-8 is known to be unnecessary for translational enhancement (Gallie *et al.*, 1987a), the loss of viability of the mutant [1/9] also may not be due to the loss of capacity for *in vivo* translation of the 130K protein gene. Although little is understood about the role of these essential sequences, the complementary sequences of these regions seem to be important to serve as a template for the plus-strand RNA synthesis.

Previously, the 5' leader sequence has been shown to bind two 80S ribosomes in the presence of sparsomycin, and the binding site for the second ribosome (in addition to the binding site at the AUG initiation codon) was mapped to an AUU (residues 15-17) (Tyc *et al.*, 1984). Moreover, the region containing this AUU has been suggested to interact with the 3' end of the wheat 18S rRNA (Yokoe *et al.*, 1983). However, a TMV mutant lacking this region was infectious and therefore we conclude that this region is dispensable for viral multiplication processes including the translation of the 130K and 180K proteins. However, deletions of longer sequences resulted in some decrease in specific infectivity, suggesting that the 5' leader sequence may serve to maintain a proper structure or spacing for efficient translation and/or RNA replication.

REFERENCES

Boccara, M., Hamilton, W.D.O., and Baulcombe, D.C. (1986). The organization and interviral homologies of genes at the 3' end of tobacco rattle virus RNA 1. *EMBO J.* **5**, 223-230.

Bujarski, J.J., Ahlquist, P., Hall, T.C., Dreher, T.W., and Kaesberg, P. (1986). Modulation of replication, aminoacylation and adenylation *in vitro* and infectivity *in vivo* of BMV RNAs containing deletions within the multifunctional 3' end. *EMBO J.* **5**, 1769-1774.

Cornelissen, B.J.C., Linthorst, H.J.M., Brederode, F.T., and Bol, J.F. (1986). Analysis of the genome structure of tobacco rattle virus strain PSG. *Nucleic Acids Res.* **14**, 2157-2169.

Culver, J.N., and Dawson, W.O. (1989). Point mutations in the coat protein gene of tobacco mosaic virus induce hypersensitivity in Nicotiana sylvestris. *Mol. Plant-Microbe Interact.* **2**, 209-213.

Dawson, W.O., Beck, D.L., Knorr, D.A., and Grantham, G.L. (1986). cDNA cloning of the complete genome of tobacco mosaic virus and production of infectious transcripts. *Proc. Natl. Acad. Sci. USA* **83**, 1832-1836.

Dawson, W.O., Bubrick, P., and Grantham, G.L. (1988). Modifications of the tobacco mosaic virus coat protein gene affecting replication, movement, and symptomatology. *Phytopathology* **78**, 783-789.

Deom, C.M., Oliver, M.J., and Beachy, R.N. (1987). The 30-kilodalton gene product of tobacco mosaic virus potentiates virus movement. *Science* **237**, 389-394.

Dorokhov, Y.L., Alexandrova, N.M., Miroshnichenko, N.A., and Atabekov, J.G. (1984). The informosome-like virus specific ribonucleoprotein (vRNP) may be involved in the transport of tobacco mosaic virus infection. *Virology* **137**, 127-134.

Dreher, T.W., and Hall, T.C. (1988). Mutational analysis of the sequence and structural requirements in brome mosaic virus for minus strand promoter activity. *J.Mol.Biol.* **201**, 31-40.

Dreher, T.W., Rao, A.L.N., and Hall, T.C. (1989). Replication *in vivo* of mutant brome mosaic virus RNAs defective in aminoacylation. *J.Mol.Biol.* **206**, 425-438.

Fraser, R.S.S. (1985). Mechanisms of Resistance to Plant Disease, Martinus Nijhoff/Junk, Dordrecht.

Gallie, D.R., Sleat, D.E., Watts, J.W., Turner, P.C., and Wilson, T.M.A. (1987a). The 5'-leader sequence of tobacco mosaic virus RNA enhances the expression of foreign gene transcripts *in vitro* and *in vivo*. Nucleic Acids. Res. **15**, 3257-3273.

Gallie, D.R., Sleat, D.E., Watts, J.W., Turner, P.C., and Wilson, T.M.A. (1987b). *In vivo* uncoating and efficient expression of foreign mRNAs packaged in TMV-like particles. *Science* **236**, 1122-1124.

Gallie, D.R., Sleat, D.E., Watts, J.W., Turner, P.C., and Wilson, T.M.A. (1988). Mutational analysis of the tobacco mosaic virus 5'-leader for altered ability to enhance translation. *Nucleic Acids Res.* **16**, 833-893.

Goelet, P., Lomonossoff, G.P., Butler, P.J.G., Akam, M.E., Gait, M.J., and Karn, J. (1982). Nucleotide sequence of tobaco mosaic virus RNA. *Proc. Natl. Acad. Sci. USA* **79**, 5818-5822.

Haseloff, J., Goelet, P., Zimmern, D., Ahlquist, P., Dasgupta, R., and Kaesberg, P. (1984). Striking similarities in amino acid sequence among nonstructural proteins encoded by RNA viruses that have dissimilar genomic organization. *Proc. Natl. Acad. Sci. USA* **81**, 4358-4362.

Hunter, T.R., Hunt, T., Knowland, J., and Zimmern, D. (1976). Messenger RNA for the coat protein of tobacco mosaic virus. *Nature* **260**, 759-764.

Ishikawa, M., Meshi, T., Motoyoshi, F., Takamatsu, N., and Okada, Y. (1986). *In vitro* mutagenesis of the putative replicase genes of tobacco mosaic virus. *Nucleic Acids Res.* **14**, 8291-8305.

Ishikawa, M., Meshi, T., Watanabe, Y., and Okada, Y. (1988). Replication of chimeric tobacco mosaic viruses which carry heterologous combinations of replicase genes and 3' noncoding regions. *Virology* **164**, 290-293.

Kamer, G., and Argos, P. (1984). Primary structural comparison of RNA-dependent polymerases from plant, animal and bacterial viruses. *Nucleic Acids Res.* **12**, 7269-7282.

Knorr, D.A., and Dawson, W.O. (1988). A point mutation in the tobacco mosaic virus capsid protein gene induces hypersensitivity in *Nicotiana sylvestris*. *Proc. Natl. Acad. Sci. USA* **85**, 170-174.

Leonard, D.A., and Zaitlin, M. (1982). A temperature-sensitive strain of tobacco mosaic virus defective in cell-to-cell movement generates an altered viral-encoded protein. *Virology* **117**, 416-424.

Meshi, T., Ishikawa, M., Motoyoshi, F., Semba, K., and Okada, Y. (1986). *In vitro* transcription of infectious RNAs from full-length cDNAs of tobacco mosaic virus. *Proc. Natl. Acad. Sci. USA* **83**, 5043-5047.

Meshi, T., Watanabe, Y., Saito, T., Sugimoto, A., Maeda, T., and Okada, Y. (1987). Function of the 30 kd protein of tobacco mosaic virus: involvement in cell-to-cell movement and dispensability for replication. *EMBO J.* **6**, 2557-2563.

Meshi, T., Motoyoshi, F., Adachi, A., Watanabe, Y., Takamatsu, N.,and Okada, Y. (1988). Two concomitant base substitutions in the putative replicase genes of tobacco mosaic virus confer the ability to overcome the effects of a tomato resistance gene, Tm-1. *EMBO J.* **7**, 1575-1581.

Meshi, T., Motoyoshi, F., Maeda, T., Yoshiwoka, S., Watanabe, H., and Okada, Y. (1989). Mutations in the tobacco mosaic virus 30 kD protein gene overcome Tm-2 resistance in tomato. *Plant Cell* **1**, 515-522.

Ohno, T., Takamatsu, N., Meshi, T., Okada, Y., Nishiguchi, N., and Kiho, T. (1983). Single amino acid substitution in 30K protein of TMV defective in virus transport function. *Virology* **131**, 255-258.

Ohno, T., Aoyagi, M., Yamanashi, Y., Saito, H., Ikawa, S., Meshi, T., and Okada, Y., (1984). Nucleotide sequence of the tobacco mosaic virus (tomato strain) genome and comparison with the common strain genome. J. Biochem. **96**, 1915-1923.

Okada, Y., Meshi, T., Watanabe, Y., Ishikawa, M., Saito, T., and Takamatsu, N. (1988). Genetic manipulation of tobacco mosaic virus. Abstracts of 5th International Congress of Plant Pathology in Kyoto, Japan.

Ooshika, I., Watanabe, Y., Meshi, T., Okada, Y., Igano, K., Inuoye, K., and Yoshida, N. (1984). Identification of the 30K protein of TMV by immunoprecipitation with antibodies directed against a synthetic peptide. *Virology* **132**, 71-78.

Rietvelt, K., Linschooten, K., Pleij, C.W.A., and Bosch, L. (1984). The three-dimensional folding of the tRNA-like structure of tobacco mosaic virus RNA. A new building principle applied twice. *EMBO J.* **3**, 2613-2619.

Rosenberg, A.H., Lade, B.N., Chui, D., Lin, S., Dunn, J.J., and Studier, F.W. (1987). Vectors for selective expression of cloned DNAs by T7 RNA polymerase. *Gene* **56**, 125-135.

Saito, T., Meshi, T., Takamatsu, N., and Okada, Y. (1987). Coat protein gene sequence of tobacco mosaic virus encodes a host response determinant. *Proc. Natl. Acad. Sci. USA* **84**, 6074-6077.

Saito, T., Imai, Y., Meshi, T., and Okada, Y. (1988). Interviral homologies of the 30K proteins of tobamoviruses. *Virology* **167**, 653-656.

Saito, T., Yamanaka, K., Watanabe, Y., Takamatsu, N., Meshi, T., and Okada, Y. (1989). Mutational analysis of the coat protein gene of tobacco mosaic virus in relation to hypersensitive response in tobacco plants with the N' gene. *Virology* **173**, 11-20.

Saito, T., Yamanaka, K., and Okada, Y. (1990). Long-distance movement and viral assembly of tobacco mosaic virus mutants. *Virology* (In press).

Siegel, A., Zaitlin, M., and Sehgal, O.P. (1962). The isolation of defective tobacco mosaic virus strains. *Proc. Natl. Acad. Sci. USA* **48**, 1845-1851.

Studier, F.W., Rosenberg, A.H., and Dunn, J.J. (1989). Use of T7 RNA polymerase to direct expression of cloned genes. *Methods in Enzymology,* (In press).

Takamatsu, N., Ohno, T., Meshi, T. and Okada, Y (1983). Molecular cloning and nucleotide sequence of the 30K and the coat protein cistron of TMV (tomato strain) genome. *Nucleic Acids Res.***11**, 3767-3778.

Takamatsu, N., Ohno, T., Meshi, T., and Okada, Y. (1987). Expression of bacterial chloramphenicol acetyltransferase gene in tobacco plants mediated by TMV-RNA. *EMBO J.* **6**, 307-311.

Tomenius, K., Clapham, D., and Meshi, T. (1987). Localization by immunogold cytochemistry of the virus-coded 30K protein in plasmodesmata of leaves infected with tobacco mosaic virus. *Virology* **160**, 363-371.

Tyc, K., Konarska, M., Gross, H.J., and Filipowicz, W. (1984). Multiple ribosome binding to the 5'-terminal leader sequence of TMV RNA. Assembly of an 80S ribosome-mRNA complex at the AUU codon. *Eur.J.Biochem.* **140**, 503-511.

Van Belkum, A., Abrahams, J.P., Pleij, C.W.A., and Bosch, L. (1985). Five pseudoknots are present at the 204 nucleotide-long 3' noncoding region of tobacco mosaic virus RNA. *Nucleic Acids Res.* **13**, 7673-7678.

Watanabe, T., Emori, Y., Ooshika, I., Meshi, T., Ohno, T., and Okada, Y. (1984). Synthesis of TMV-specific RNAs and proteins at the early stage of infection in tobacco protoplasts: Transient expression of the 30K protein and its mRNA. *Virology* **133**, 18-24.

Watanabe, T., Ooshika, I., Meshi, T., and Okada, Y. (1986). Subcellular localization of the 30K protein in TMV-inoculated tobacco protoplasts. *Virology* **152**, 414-420.

Watanabe, T., Meshi, T., and Okada, Y. (1987). Infection of tobacco protoplasts with *in vitro* transcribed tobacco mosaic virus RNA using an improved electroporation method. *FEBS Lett.* **219**, 65-69.

Yokoe, S., Tanaka, M., Hibasami, H., Nagai, J., and Nakashima, K. (1983). Cross-linking of tobacco mosaic virus RNA and capped polyribonucleotides to 18S rRNA in wheat germ ribosome-mRNA complexes. *J.Biochem.* **94**, 1803-1808.

Zimmern, D., and Hunter, T., (1983). Point mutation in the 30K open reading frame of TMV implicated in temperature-sensitive assembly and local lesion spreading of mutant Ni 2519. *EMBO J.* **2**, 1893-1900.

DISCUSSION OF Y. OKADA'S PRESENTATION

R. Goldbach: You made some deletions in the pseudoknot region that led to tolerant viruses not showing symptoms. Did you do some cross protection experiments with those strains?

Y. Okada: Yes, I have done those experiments. The plants show cross protection, but I only used three plants, and I am thus not sure if they have a strong ability for cross protection. I want to do it again with many plants, but these experiments are restricted to greenhouses in Japan, so I can only do it on a small scale.

R. Hull: The localization of the 30K protein in plants has previously been done with the oligopeptide serum to the C-terminus, and it has been shown that the 30K gets expressed transiently. Have you done experiments using your 30K antiserum to the whole protein to see whether the transient appearance in plants is just due to the C-terminus being removed or to the whole protein being removed?

Y. Okada: I have not done it yet.

R. Beachy: Mike Deom indicates that in fact it is not transient. With the antibodies that we use, the protein continues to accumulate in the wall fraction. It may be transient in the free state in the cytoplasm, but it accumulates in the wall to quite a stable level.

J. Atabekov: Concerning your non-assembling mutant, can you rule out another alternative explanation, for example that informosome-like particles move, but nevertheless the origin of assembly could be essential for viral ribonucleoprotein assembly?

Y. Okada: I cannot exclude the possibility that the long distance movement is caused by the ribonucleoprotein complex you suggested, but in that case the formation of the ribonucleoprotein complex may also be necessary. We have tried several times to isolate the ribonucleoprotein complex in our system, but we did not find such a particle yet. But our method is not the same as yours, so we cannot exclude the possibility completely.

J. Atabekov: We sent the experimental protocol to Dr. Meishi.

P. Palukaitis: Have you looked at the yellow spots produced in the upper leaves, reinoculated those to plants, and then do you get the same kind of infection - very poor with a few yellow spots, or do you get now a normal systemic infection? In other words has there been some mutation of the virus particle, and what you've measured is a low level of virus mutation which has become localized.

Y. Okada: If you extract the chlorotic spot, it has no infectivity. The RNA is not assembled, and by extraction it is digested.

P. Palukaitis: Yes, but you have found a very low level of particles on the gradient by electron microscopy.

Y. Okada: I found a low level of particles in the inoculated leaves, but we cannot find particles in the upper, uninoculated leaves.

P. Palukaitis: Do you think you are getting encapsidation at the site in the coat protein that is sort of a primitive origin of encapsidation, like the one in the cowpea strain at that position, in the case where you see some particles?

Y. Okada: I have not done this experiment so I cannot say.

P. Palukaitis: With the deletion in the pseudoknot region that produces an attenuated virus, does it still produce local lesions on N gene-containing tobacco?

Y. Okada: Yes.

M. Zaitlin: Some years ago your lab reported mutations in the 183K protein *in vitro*. Have you done any more work with that to seek the function of the 183K protein in TMVreplication?

Y. Okada: No.

Relationship of Tobacco Mosaic Virus Gene Expression to Movement Within the Plant

William O. Dawson

Department of Plant Pathology and Genetics Graduate Group
University of California, Riverside, CA 92521,USA

Tobacco mosaic virus (TMV) is the type member of the tobamovirus group that is characterized by 18 X 300 nm virions made up of about 2000 units of a single structural protein and one molecule of plus-sense RNA of approximately 6400 nucleotides. TMV has a wide host range and occurs throughout the world causing substantial crop losses. The virus replicates in almost every cell of the plant and accumulates to unusually high titers. Disease generally is caused by abnormal chloroplast development in infected plants resulting in leaves that are deficient in photosynthetic capacity and exhibit typical "mosaic" symptoms of light and dark green leaves.

The genome of TMV has four large open reading frames (ORFs) (Goelet *et al.*, 1982) (Fig. 1). The 5' proximal ORF and a readthrough ORF are translated from genomic RNA into 126K and 183K proteins (Pelham, 1978) that are required for replication (Ishikawa *et al.*, 1986). The 3' proximal genes encode the 30K and capsid (17.5 kd) proteins that are expressed through 3'-coterminal subgenomic mRNAs (Hunter *et al.*, 1976; Siegel *et al.*, 1976; Beachy and Zaitlin, 1977).

In nature, plants normally become infected with TMV at a single wound site. Infection is established in the initially infected cell after which the virus moves to the rest of the plant, continuing to move into newly developing parts of the plant as it grows. Often almost all cells of the plant, with the exception of meristematic and gametic cells, become infected with this virus. The spread of the infection throughout the plant can be divided into two functional mechanisms, cell-to-cell movement and long-distance movement.

Cell-to-cell movement allows transport of the infectious agent between adjacent cells within and between tissues. These cells are separated by cell walls. The virus is thought to move through plasmodesmata that provide protoplasmic connections between adjacent cells. Cell-to-cell movement is generally responsible for transport within leaves and other plant organs. The rate of movement is in the order of micrometers per hour.

Spread of the viral infectious agent from leaf to leaf and throughout distal parts of the plant is by long-distance movement. This transport occurs through phloem tissue and movement occurs most prevalently in the direction of movement of metabolites in the phloem, that is, movement is most rapid towards tissues importing metabolites, young shoots and roots in contrast to mature leaves. The rate of movement in phloem can be in the order of centimeters per hour.

Until a few years ago, the dogma was that viruses moved passively. Plasmodesmata were considered to be inert channels between cells through which viruses slowly moved until they moved into the phloem where they were quickly dispersed through the plant. There now is evidence that both of these processes require virus-specific functions.

CELL-TO-CELL MOVEMENT

The viral 30K protein is involved in cell-to-cell movement

The first demonstration that a virus-specific function was required for cell-to-cell movement was by Nishiguchi *et al.* (1978), who isolated a TMV mutant that at the restrictive temperature was able to replicate efficiently but was not able to spread from cell to cell. This demonstrated that a virus-specific function was required, but it was not known what viral gene product was involved. Leonard and Zaitlin (1982) then showed that the 30K protein of this mutant had an alteration and could be the gene product involved in the movement function. The definitive demonstrations of the requirement of the 30K gene for cell-to-cell movement were by Meshi *et al.* (1987) who made directed mutations in the 30K gene resulting in virus that replicated in protoplasts but could not move in plants and by Deom *et al.* (1987) who showed that movement-defective mutants could move in transgenic plants expressing the 30K gene.

With TMV, the coat protein is not involved in cell-to-cell movement. Mutants with the entire coat protein gene deleted move cell to cell identically to the wild type virus (Dawson *et al.*, 1988). Thus the virus can efficiently move from cell to cell in some form other that as virions. However, this does not appear to be a universal phenomenon among plant viruses. Viruses of the bromo- (French *et al.*, 1986) and comovirus groups (Wellink and van Kammen, 1989) appear to need a functional coat protein gene for cell-to-cell movement.

The 30K protein is associated with the cell wall

Although the 30K protein has been reported to accumulate in nuclei of infected protoplasts (Watanabe *et al.*, 1986), most evidence suggests that in intact leaves the 30K protein is associated with the cell wall. *In situ* localization demonstrated that in intact leaves the 30K protein accumulated in plasmodesmatal areas of the cell wall (Tomelius *et al.*, 1987). Similar movement proteins of other plant viruses also have been localized in the cell wall areas by *in situ* labelling (Stussi-Garaud *et al.*, 1987; Linstead *et al.*, 1988). Procedures that have effectively extracted the movement proteins were designed to extract proteins that were tightly associated with the non-soluble cell wall fraction (Godefroy-Colburn *et al.*, 1986). In a recent subcellular fractionation study, Moser *et al.* (1988) confirmed the association of the 30K protein with the non-soluble cell wall fraction, and additionally, transiently detected the protein in the cytoplasmic membrane fraction. The localization of the protein in cell walls suggests that in intact tissue it is actively transported out of the cytoplasm.

The 30K protein is produced transiently during the early infection period

The 30K gene is expressed only during the early period of the infection of a cell. In protoplasts, the 30K protein was synthesized transiently between 2 and 10 hours after infection, whereas the other viral proteins continued to be produced (Watanabe *et al.*, 1984; Blum *et al.*, 1989). Also, this protein was produced only in minute amounts. However, the amounts of 30K protein found in intact cells (Moser *et al.*, 1988; Lehto *et al.*, 1990 a) are substantially higher than those produced in protoplasts. We confirmed that most of the 30K protein was produced during the early phase of TMV replication in intact tobacco cells as was reported in studies with tobacco protoplasts (Lehto *et al.*, 1990 a). However, our observation of the accumulation of 30K protein differed from that detected in protoplasts. In intact cells the protein continued to accumulate much longer than in protoplasts, until about 36 hours of replication. Some type of regulation apparently represses 30K synthesis in protoplasts. Also, the 30K protein did not disappear after the early phase of the infection as it does in protoplasts, but remained detectable throughout the sampling periods, as late as 24 days after inoculation, similar to other reports of 30K protein in leaves (Moser *et al.*, 1988). These data suggest that production and disposition of this protein may be different in protoplasts than in intact cells. The difference in levels of this protein in intact cells compared to protoplasts could be due to the protein being transported out of the protoplasts and perhaps degraded. These data demonstrate that accumulation and possibly regulation of the 30K protein is different in cells of intact leaves than in protoplasts. In intact cells, 30K protein is stably deposited in the cell wall. The export of the protein out of the cytoplasm may cause it to be synthesized for longer periods of time in intact leaves compared to protoplasts. The lack of deposition of the protein in cell walls might lead to its rapid accumulation and shut-down of synthesis.

Another observation concerning the regulation of synthesis of the 30K protein is the effect of actinomycin D-treatment. Actinomycin D has been shown to selectively enhance the synthesis of this protein up to 100-fold in protoplasts (Blum *et al.*, 1989). Although the mechanism of this enhancement is not understood, these data suggest that actinomycin D affects the regulation of the 30K gene. It is possible that actinomycin D stimulates some product that is required for continued 30K protein synthesis or interferes with a product required to turn off 30K synthesis.

Additionally, this observation should be a important consideration in designing experiments to examine regulation of 30K protein synthesis. Since viral infections of plants cause little inhibition of host RNA synthesis, actinomycin D routinely has been used to suppress host RNA synthesis when examining viral RNA synthesis. However, since this compound may alter the synthesis of 30K mRNA, this makes monitoring 30K mRNA synthesis more difficult.

Effects of different amounts of 30K protein on cell-to-cell movement

Reduced production of 30K protein. We have made a number of different TMV mutants with alterations of the viral genome to examine regulation of gene expression and effects of genome structure on virus replication and stability of the altered mutants during replication (Dawson *et al.*, 1988; Lehto and Dawson, 1990 a; Lehto *et al.*, 1990 b; Lehto and Dawson, 1990 b; Dawson *et al.*, 1989; Lehto, Culver, and Dawson, unpublished). Although these mutants were examined for other reasons, several of the mutants produced altered amounts of 30K protein and resulted in alterations in movement functions. Here, I will attempt to summarize the relationship of 30K production to ability of the mutants to move from cell to cell.

Mutant KK1. A mutant with a seven nucleotide insertion in front of the 30K ORF to alter the start codon sequence context (Fig. 1) was infectious and replicated indistinguishably from wild type TMV when identical plants were inoculated and compared:

wild type CUUUAUAGAUGGCUCUAG

KK1 GUUUAUAGACUCGAGAUGGCUCUAG

However, mutant KK1 produced only 30% as much 30K protein as did wild type TMV (Lehto and Dawson, 1990 a). There was no reduction of production of other viral proteins or viral RNAs. Although mutant KK1 accumulated substantially reduced amounts of 30K protein, cell-to-cell spread appeared equal to that of wild type TMV, estimated by local lesion size and the appearance of local and systemic symptoms at the same time as those in wild type TMV infections. This suggests that the reduced level of 30K protein in KK1 infections was adequate to mediate normal viral spread.

Mutant CP30K. Another mutant had the entire 30K ORF fused in frame behind approximately two-thirds of the coat protein reading frame (Fig. 1). The resulting virus, CP30K, was a free-RNA virus because of the lack of an intact capsid protein. This resulted in production of relatively large amounts of the fusion protein, but reduced amounts of native 30K protein (Lehto and Dawson, 1990 b). The level of 30K protein found in infected cells was only at the lower level of detection, approximately 10-50 times less than that found in wild type TMV-infected cells. However, with this reduction in accumulation of 30K protein, there was only a slight reduction in the ability of this mutant to move cell to cell. In the local lesion host, Xanthi nc tobacco, this mutant induced lesions that developed two days later (4-5 days) than those produced by wild type TMV. The CP30K lesions were approximately one-half the size of those induced by wild type TMV. Even at this reduced level of 30K protein, cell-to-cell movement was only slightly impaired.

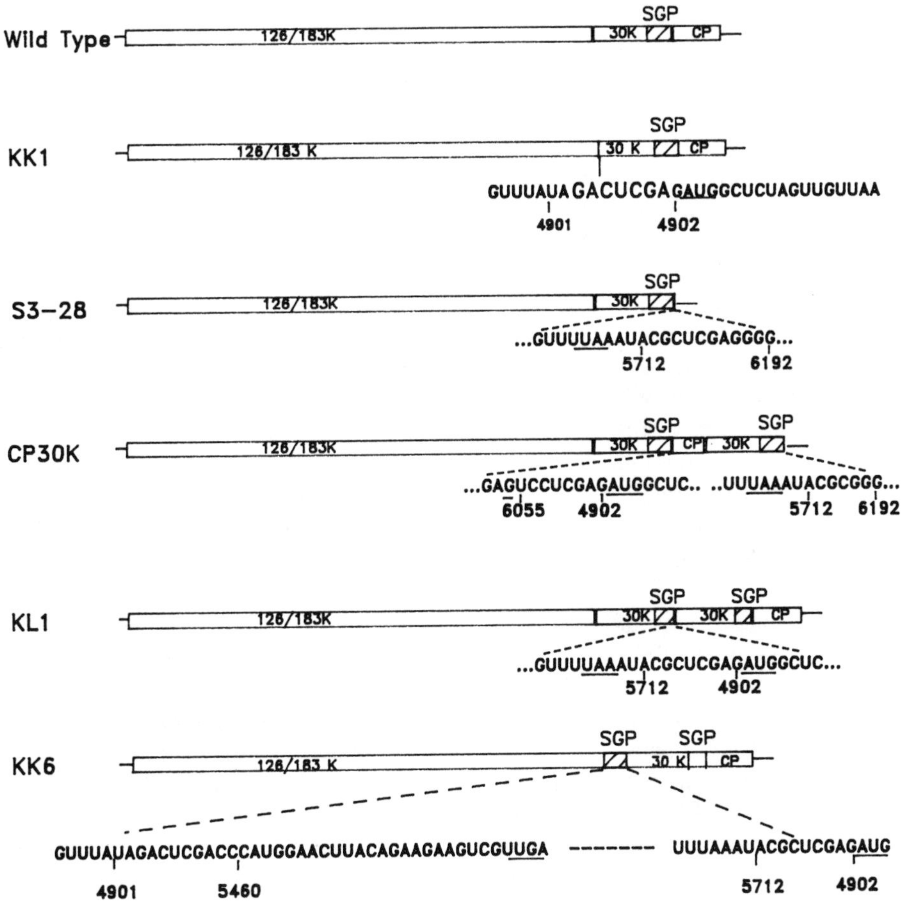

Fig. 1 *In vitro* constructed mutants of TMV. Numbers are nucleotides of wild type (strain U1) TMV according to Goelet *et al.* (1982). Start and stop codons are underlined. Reading frames are boxed. The coat protein subgenomic mRNA promoter, which is the 3' region of the 30K gene, is labeled SGP.

Mutant KL1. Mutant KL1 had a second 30K gene inserted downstream of the native 30K gene, which positioned it behind the coat protein subgenomic RNA promoter (Fig. 1). The upstream 30K gene was in its normal position relative to its subgenomic RNA promoter and the 5' proximal genes, and the sequences down stream of the insertion contained the coat protein gene in its normal position relative to a second coat protein subgenomic RNA promoter and to the 3' non-coding region.

The level of replication of mutant KL1 was similar to that of wild type TMV (Lehto and Dawson, 1990 b). The amounts of genomic and coat protein subgenomic mRNA were similar between the two viruses, and yields of purified virions and amounts of infectivity from KL1-infected leaves were similar to those from wild type TMV. When the accumulation of 30K, 126K, and coat proteins in inoculated and systemically infected leaves of plants infected with KL1 for different periods of time was assayed and compared to that of wild type TMV, the accumulation of 126K and coat proteins was only slightly reduced, indicating that the mutant was not appreciably defective in ability to replicate and systemically invade intact plants. However, 30K protein was barely detectable, between one-tenth and one-hundredth that of wild type TMV.

Cell-to-cell movement of this mutant was only slightly less than that of wild type TMV in spite of the dramatic decrease in 30K protein. Local lesions developed in Xanthi nc plants in 2-3 days. When opposite half-leaves of Xanthi nc were inoculated with KL1 and wild type TMV, KL1 local lesions were approximately one-half the diameter of those induced by wild type TMV, suggesting that this mutant was reduced in ability to move from cell to cell and/or ability to replicate. Since KL1 appears to replicate efficiently, this suggests that the reduction in size of lesions was due primarily to reduced ability of mutant KL1 to move cell to cell.

Increased production of 30K protein. Positioning of the 30K gene nearer to the 3' terminus by deletion of portions of the coat protein gene proportionally increased the amount of 30K protein produced (Lehto, Culver, and Dawson, in preparation). Mutants with the coat protein gene completely deleted, S3-28 (Fig. 1), produced approximately 10-50 times more 30K protein than the wild type virus. Mutants with deletions in the coat protein gene also had increased production of 30K protein directly proportional to the number of nucleotides removed. Production of 126K protein by these mutants was the same as that of the parental wild type virus. However, none of the mutants moved from cell to cell faster that did wild type TMV. Lesions formed in Xanthi nc tobacco were all identical to those of the wild type virus.

Overall, results from these mutants suggest that there is a threshold amount of 30K protein, which appears to be approximately one-tenth that produced by wild type TMV, that is sufficient for maximal cell-to-cell movement within the plant. Amounts over this threshold have no effect. Amounts less than the threshold level appear to result in gradual decreases in movement.

Effects of timing of 30K protein on movement

Mutant KK6 contained a 250 nucleotide fragment containing the coat protein subgenomic mRNA promoter and leader inserted in front of the 30K ORF (Fig. 1). This put the 30K ORF under the regulation of the coat protein subgenomic mRNA promoter and leader (Lehto *et al.*, 1990 b). Instead of the normal 75 nucleotide leader, the KK6 30K mRNA had a hybrid coat protein mRNA leader of 24 nucleotides. The time course of accumulation of 30K protein by KK6, in contrast to the other KK6 proteins, was greatly delayed compared to that of wild type TMV. In Xanthi tobacco leaves mechanically inoculated with wild type TMV, 30K protein always accumulated to maximal levels (2-3 days after inoculation) before 126K protein (3-5 days after inoculation). In contrast, the 126K protein of mutant KK6 always increased to maximal levels earlier (4-5 days after inoculation) than KK6 30K protein (6-10 days after inoculation). Mutant KK6 30K protein accumulated 4-7 days later than 126K protein instead of 1-2 days before the 126K protein as in wild type TMV infections. The final concentration of 30K protein in KK6-infected leaves in different experiments was approximately equal to or slightly greater than the maximal levels in wild type TMV-infected leaves. However, this level was attained after 6-12 days in KK6-infected leaves compared to 2-3 days in wild type TMV-infected leaves.

Mutant KK6 produced lesions in Xanthi nc tobacco that developed more slowly (5 days) than lesions induced by wild type TMV (2-3 days). The necrotic lesions induced by KK6 were much smaller, approximately 10%-20% the diameter of those caused by wild type TMV. Defective movement could be due to the delayed production of the 30K protein. However, the 183K gene of KK6 also was altered, which could cause replication to be reduced. As discussed earlier, we have shown that the production of 30K protein can be greatly increased by removing the coat protein gene. To examine whether the delayed 30K protein accumulation and/or the altered 183K protein were responsible for the small lesion phenotype of KK6, the coat protein gene of mutant KK6 was removed, resulting in mutant KK8 which retained the coat protein subgenomic mRNA promoter inserted in the 183K ORF.

Mutant KK8 replicated in Xanthi and Xanthi nc tobacco, resulting in a free-RNA-virus infection with no detectable coat protein produced. Approximately 20 times as much 30K protein accumulated in KK8-infected leaves as in KK6- or wild type TMV-infected leaves. Lesions produced by KK8 were substantially larger than those of KK6, and almost the same size and produced at the same time after inoculation as those of wild type TMV. This suggests that the small local lesion phenotype of KK6 was due to reduced ability to move cell to cell caused by the delayed production of the 30K protein and not to impaired replication due to alterations of the 183K gene.

The altered kinetics of 30K protein accumulation by mutant KK6 affected the virus phenotype. The final level of mutant KK6 30K protein was approximately the same as that of wild type TMV, but it occurred late instead of early in the infection. When the production of 30K protein was increased in mutant KK8, the defect in cell-to-cell movement was overcome although the 183K gene remained altered. The other mutants described above, with reduced production of 30K protein, resulted in only small reductions in the ability to move cell to cell, whereas mutant KK6 had more restricted movement. This suggests that the movement defect of mutant KK6 was not due to too little 30K, but was due to the delayed production of 30K protein. The 30K protein apparently is needed during the early hours of the replication cycle to properly mediate cell-to-cell transport and 30K protein produced later does not function adequately.

LONG-DISTANCE MOVEMENT

There is little information concerning how viruses move over long distances, but, in contrast to what has been dogma, it is likely that virus-specific functions also are required for this process. There are numerous examples of viruses that move from cell to cell within an inoculated leaf, but are not able to spread to other parts of the plant. We are examining an artificially constructed hybrid virus made up of the common and orchid strains of TMV that has similar movement properties. This hybrid virus moves efficiently from cell to cell in inoculated leaves but is unable to move to other leaves of the plant (Hilf and Dawson, unpublished). Substitution of orchid strain sequences into the U1 strain created a virus that replicated efficiently and moved cell to cell, but was unable to move long distances. These data suggest that virus specific functions are required for this process.

Effects of coat protein on long-distance movement

One likely candidate for the virus-specific function is the coat protein. It has been known that virus mutants without a functional coat protein are greatly deficient in long-distance movement (Siegel *et al.*, 1962). However, these mutants do not assemble into virions. One possibility is that the function required of the coat protein for efficient long-distance movement is formation of virions. The infectious agent that can move from cell to cell in a non-virion form may not survive long-distance movement. However, the TMV U1-orchid strain hybrid that is deficient in long-distance movement forms stable virions, suggesting that formation of virions is not the key to long-distance movement. Recently, Y. Okada reported that mutants of TMV that produced normal coat protein, but were mutated so that the RNA could not assemble into virions, could move efficiently over long distances (presented at 5th International Congress of Plant Pathology, Kyoto, Japan, Aug. 20-27, 1988). We have shown that mutants of TMV with different sized deletions in the coat protein gene differ greatly in their ability to move long distances (Dawson *et al.*, 1988). None of the mutants assemble into virions. However, some of the mutants move into almost all of the leaves of developing plants whereas others rarely move out of the inoculated leaves. These data suggest that the coat protein is involved in long-distance movement in some manner other than the formation of virions. This may also be related to the prevention of movement of TMV through stem segments from coat protein transgenic plants when a leaf is attached to the stem segment grafted into tobacco plants (R. Beachy, presented at 5th International Congress of Plant Pathology, Kyoto, Japan, Aug. 20-27, 1988). It is possible that the coat protein is required for phloem loading of an infectious virus complex.

The requirement for coat protein is not absolute. Mutants that produce no coat protein, either because the start codon of the coat protein gene has been mutated or because the entire gene has been removed, eventually can move into upper parts of the plant (Culver and Dawson, in press). Weeks after inoculation of lower leaves of *N. sylvestris*, some plants develop normal mosaic symptoms in upper leaves. No coat protein was detected in the leaves with mosaic symptoms demonstrating that the coat protein was not involved in symptom production and that long-distance movement could occur in the absence of coat protein. However, this movement was substantially less efficient than that of virus forming virions.

Effects of 30K protein on long-distance movement

Since the coat protein is involved in long-distance movement, only mutants that produce normal amounts of coat protein and assemble into virions can be examined relative to their ability to move long distances. Mutants KK1, KL1, and KK6 have these characteristics along with abnormal production of 30K protein.

Mutants KK1 and KL1 produce reduced amounts of 30K protein and mutant KL1 has reduced ability to move cell to cell. However, neither mutant is deficient in long-distance movement. Both induce full systemic infections in upper parts of infected plants. Thus the reduced production of 30K protein appears to have no effect on long-distance movement.

In contrast, mutant KK6, with delayed production of 30K protein, was deficient also in long-distance movement. KK6 systemically infected Xanthi plants, but the first symptoms developed slowly (14-20 days) compared to those induced by wild type TMV (about 4 days). The symptoms in the upper leaves of KK6-infected plants were distinct, isolated yellow spots, 0.2-1.0 cm in diameter, compared to the mosaic symptoms produced by wild type TMV. KK6 appeared to be localized within the yellow areas of systemically infected leaves based on infectivity assays. It is not clear whether this defect results from a defect in long-distance movement or whether the apparent defect in long distance movement is due to the lack of cell-to-cell movement after the virus moves into the upper leaves. If the second possibility is true, it would suggest that long distance movement only provides a few infection sites, because many upper leaves only had 10-30 yellow spots.

Effects of host on long-distance movement

In contrast to the greatly reduced ability of mutant KK6 to systemically infect tobacco and several other hosts examined, mutant KK6 was not defective in movement in *Nicotiana benthamiana*. In this plant, the virus causes a vein-clearing symptom in upper leaves 3-4 days after inoculation of the lower leaves, the same as occurs in wild type TMV-infected plants. Assays of virus in the upper leaves after 5-6 days resulted in approximately the same amount of infectivity, although the virus from KK6-infected plants induced typical small lesions in Xanthi nc tobacco. This suggests that the movement defect of mutant KK6 does not occur in *N. benthamiana*. This is the only host in which mutant KK6 was found to move efficiently.

N. benthamiana is an often used laboratory host of a wide range of viruses for propagation to high titers. In some cases, viruses appear to spread more effectively in this plant than in other hosts (Rushing *et al.*, 1987). Does this suggest that this plant is different from other plants in its movement requirements?

SUMMARY

Cell-to-cell movement and probably long-distance movement within the plant are processes that require active participation by the virus. A movement protein (and the coat protein for some viruses) is required for cell-to-cell movement. The timing of production of this protein appears to be more important than the amount of this protein produced. Long-distance movement appears to be as complex as cell-to-cell movement, but this process remains to be resolved. One of the more interesting aspects of these processes is the hosts' contributions, of which we have almost no information. Understanding these processes could lead to new approaches to controlling virus diseases.

ACKNOWLEDGEMENTS

This research was supported in part by grants from the National Science Foundation (DMB 8607638), US Department of Agriculture (8900363), and US- Israel BARD (I-882-85).

REFERENCES

Beachy, R. N., and Zaitlin, M. (1977). Characterization and *in vitro* translation of the RNAs from less-than-full-length, virus related, nucleoprotein rods present in tobacco mosaic virus preparations. *Virology* **81**, 160-169.

Blum, H., Gross, H. J., and Beier, H. (1989). The expression of the TMV- specific 30-kDa protein is strongly and selectively enhanced by Actinomycin. *Virology* **169**, 51-61.

Dawson, W. O., Bubrick, P., and Grantham, G. L. (1988). Modifications of thetobacco mosaic virus coat protein gene affecting replication, movement, and symptomatology. *Phytopathology* **78**, 783-789.

Dawson, W. O., Lewandowski, M., Hilf, M. E., Bubrick, P., Raffo, A. J., Shaw, J. J., Grantham, G. L., and Desjardins, P. R. (1989). A tobacco mosaic virus-hybrid expresses and loses and added gene. *Virology* **172**, 285-292.

Deom, C. M., Oliver, M. J., and Beachy, R. N. (1987). The 30-kilodalton gene product of tobacco mosaic virus potentiates virus movement. *Science* **237**, 389-393.

Culver, J. N., and Dawson, W. O. (1989). Tobacco mosaic virus coat protein: an elicitor of the hypersensitive reaction but not required for the delevopment of mosaic symptoms in Nicotiana sylvestris. *Virology* **173**, 755-758.

French, R., Janda, M., and Ahlquist, P. (1986). Bacterial gene inserted in an engineered RNA virus: efficient expression in monocotyledenous plant cells. *Science* **231**, 1294-1297.

Godefroy-Colburn, T., Gagey, M., Berna, A., and Stussi-Garaud, C. (1986). A non-structural protein of alfalfa mosaic virus in the walls of infected tobacco cells. *J. Gen. Virol.* **67**, 2233-2239.

Hunter, T., Hunt, T., Knowland, J., and Zimmern, D. (1976). Messenger RNA for the coat protein of tobacco mosaic virus. *Nature* (London) **260**, 759-764.

Ishikawa, M., Meshi, T., Motoyoshi, F., Takamatsu, N., and Okada, Y. (1986). *In vitro* mutagenesis of the putative replicase genes of tobacco mosaic virus. *Nucl. Acids Res.* **14**, 8291-8305.

Lehto, K., Bubrick, P., and Dawson, W. O. (1990 a). Time course of TMV 30K protein accumulation in intact leaves. *Virology* **174**, 290-293.

Lehto, K., and Dawson, W. O. (1990 a). Changing the start codon context of the 30K gene of tobacco mosaic virus from "weak" to "strong" does not increase expression. *Virology* **173**, 169-174.

Lehto, K., and Dawson, W. O. (1990 b). Replication, stability, and gene expression of tobacco mosaic virus mutants with a second 30K ORF. *Virology*, in press .

Lehto, K., Grantham, G. L., and Dawson, W. O. (1990 b). Insertion of sequences containing the coat protein subgenomic RNA promoter and leader in front of the tobacco mosaic virus 30K ORF delays its expression and causes defective cell-to-cell movement. *Virology* **174**, 145-157.

Leonard, D. A., and Zaitlin, M. (1982). A temperature-sensitive strain of tobacco mosaic virus defective in cell-to-cell movement generates an altered viral-coded protein. *Virology* **117**, 416-424.

Linstead, P. J., Hills, G. J., Plaskitt, K. A., Wilson, I. G., Harker, C. L. and Maule, A. J. (1988). The subcellular location of the gene 1 product of cauliflower mosaic virus is consistent with a function associated with virus spread. *J. Gen. Virol.* **69**, 1809-1818.

Meshi, T., Watanabe, Y., Saito, T., Sugimoto, A., Maeda, T., and Okada, Y. (1987). Function of the 30kd protein of tobacco mosaic virus: involvement in cell-to-cell movement and dispensability for replication. *EMBO J.* **6**, 2557-2563.

Moser, O., Gagey, M. -J., Godfroy-Colburn, T., Ellwart-Tschurtz, M., Nitschko, H., and Mundry, K. -W. (1988). The fate of the transport protein of tobacco mosaic virus in systemic and hypersensitive tobacco hosts. *J. Gen. Virol.* **69**, 1367-1373.

Nishiguchi, M., Motoyoshi, F., and Oshima, N. (1978). Behaviour of a temperature sensitive strain of tobacco mosaic virus: Its behaviour in tomato leaf epidermis. *J. Gen. Virol.* **39**, 53-61.

Pelham, H. R. B. (1978). Leaky UAG termination codon in tobacco mosaic virus. *Nature* (London) **272**, 469-471.

Rushing, A. E., Sunter, G., Gardiner, W. E., Dute, R. R., and Bisaro, D. M. (1987). Ultrastructural aspects of tomato golden mosaic virus infection in tobacco. *Phytopatology* **77**, 1231-1236.

Siegel, A., Montgomery, V. H. I., and Kolacz, K. (1976). A messenger RNA for coat protein isolated from tobacco mosaic virus infected tissue. *Virology* **73**, 363-371.

Siegel, A., Zaitlin, M., and Sehgal, O. P. (1962). The isolation of defective tobacco mosaic virus strains. *Proc. Natl. Acad. Sci. USA* **48**, 1845-1851.

Stussi-Garaud, C., Garaud, J., Berna, A., and Godefroy-Colburn, T. (1987). In situ localization of an alfalfa mosaic virus non-structural protein in plant cell walls: correlation with virus transport. *J. gen. Virol.* **68**, 1779-1784.

Watanabe, Y., Emori, Y., Ooshika, I., Meshi, T., Ohno, T., and Okada, Y. (1984). Synthesis of TMV specific RNAs and proteins at the early stage of infection in tobacco protoplasts: transient expression of the 30K protein and its mRNA. *Virology* **133**, 18-24.

Watanabe, Y., Ooshika, I., Meshi, T., and Okada, Y. (1986). Subcellular localization of the 30K protein in TMV-inoculated tobacco protoplasts. *Virology* **152**, 414-420.

Wellink, J., and van Kammen, A. (1989). Cell-to-cell transport of cowpea mosaic virus requires both 58K/48K proteins and the capsid proteins. *J. Gen. Virol.* **70**, 2279-2286.

DISCUSSION OF W. DAWSON'S PRESENTATION

T. Hohn: In the case of your mutants that have a very long delay in symptom appearance do you get revertants?

W. Dawson: It depends totally on the mutants. I would say with TMV you always get revertants, but you get them at an extraordinary low level, and in fact, the one lesion that I showed you on that leaf was a revertant that had come up in it, but there were probably hundreds of infections. KK6, which has a coat protein subgenomic promoter, is one of the most stable mutants which we have developed. You can multiply that virus in tobacco for months and not have it become wild type virus. That's not true of a lot of the others.

P. Palukaitis: Have you done the last experiment you described in *Nicotiana clevelandii*?

W. Dawson: No.

P. Palukaitis: Because both *N. benthamiana* and *N. clevelandii* are much more susceptible to a larger range of viruses, I just wondered.

W. Dawson: That was one of the initial things that I was thinking about with *N. benthamiana*; in fact, I don't know if the gemini people can speak to this, but there is some suggestion that some of the geminiviruses move much better in *N. benthamiana* than maybe they do in other plants. Somebody else can talk about that better than I can.

P. Palukaitis: The only correlation we have found is that both *N. clevelandii* and *N. benthamiana* have much lower levels of the RNA-dependent RNA polymerase than other tobaccos.

P. Ahlquist: The mutant that had the coat protein subgenomic promoter in front of the 30K gene, was that KK6? (*W. Dawson*: Right.) It was probably on your slide, but was that done by insertion in such a way that the 183K readthrough domain was not altered in sequence?

W. Dawson: The end of the 183K protein's last 4 or 5 amino acids are altered and it seems to have no effect. What we have done to demonstrate that the 183K alteration is not the defect in movement is to delete the coat protein gene of that same mutant so that you increase the amount of 30K production.

The phenotype goes away and comes back and becomes almost wild type; it is no longer deficient in cell-to-cell movement.

P. Ahlquist: We find that *N. benthamiana* also supports the systemic infection of a number of the bromoviruses, and in that case what we see is that the spectrum of genes that are allowable, and the combination of genes that are allowable to support systemic spread is broader than in other hosts, but at the same time you do require intact genes for all of the various viral proteins to get systemic infection. So there doesn't seem to be a complete relaxation of any specific functional requirement.

W. Dawson: In this mutant it is not totally deficient — my guess is that if you totally delete the 30K protein it would not replicate in *N. benthamiana*, but I haven't tried that.

R. Hull: For long distance spread the virus has to move into the vascular system in three stages. It has to move into the sieve tubes, along the sieve tubes, and back out again from the sieve tubes. So obviously there are problems especially getting back out again. With KK6 have you looked to see whether the virus has actually moved up the vascular system and is not spreading into the young leaves?

W. Dawson: No.

W. Gerlach: The KK6 was the one that you did the time course with. Have you done a similar type of experiment with *N. benthamiana*, and is it accumulating much more quickly?

W. Dawson: No. We thought we saw this a year ago, but I just went back and confirmed it in the last 2 or 3 weeks. The one thing we were concerned about was whether we were getting reversion and there was some sort of selection for revertants in *N. benthamiana*, and we were getting wild type virus. It has only been recently that we found out that is not what happened.

M. Wilson: When you got the mosaic after a couple of months from the mutant truly without coat protein, could you isolate infectious RNA from both the light and the dark green islands of those leaves, and were they superinfectable?

W. Dawson: We never tried that.

M. Zaitlin: I would like to make a comment related to that. The mutants that Al Siegel, Om Seghal and I developed in the 60's — I kept those going for years; they were truly coat protein-defective and would not move long distances. However, occasionally you would see the virus get into the vascular system, and if the plant were young it got into the apex, so you would have what appeared to be a real systemic infection, but it was not. If you isolated the virus from those leaves, it was still defective, so there is this word of caution with respect to long distance movement. I think that occasionally even though those viruses were coat protein-defective, virus could get out into the vascular system.

W. Dawson: But certainly I think the important thing in long distance movement is rate, and the ones without coat protein certainly do not move anything close to the rate of wild type.

C. Hiruki: With regard to a movement into the vascular system, I think we shouldn't forget about the effect of temperature. Quite often the virus has a strain that localizes in certain species but which will easily become systemic under elevated temperature. This is well known. Sometimes this occurs only erratically; it happens only on a few occasions. Perhaps in that case you might look at the temperature record.

W. Dawson: We haven't looked at any effects of temperature on that movement yet.

A. Nejidat: When you talk about systemic spread, there is also some directional movement of the virus. Do you see any domains of the virus genome which may direct the virus in different orientations, systemic or to the roots?

W. Dawson: No, I think that's way ahead of where we are.

B. Harrison: In relation to this long distance spread, Milt Zaitlin brought up the question of his mutants of the 1960s. The same is true of tobacco rattle virus. I think you can get what appears to be a systemic movement through cell-to-cell spread in these very long periods, and the fact that you get symptoms in the tip leaves after a long time doesn't necessarily mean that the infective agent has passed through the vascular system. It can have cell-to-cell movement up the stem and I am sure that happens certainly with tobacco rattle virus, and I suspect in the mutants that Milt has mentioned. So over these very long periods I think it could be the old cell-to-cell business coming into play again.

W. Dawson: That would make it cleaner because that would say that you need coat protein for long distance movement, and it wouldn't be such a division of rates. It would be nice if it were more absolute.

The Cell to Cell Movement of Viruses in Plants

J.G. Atabekov, M.E. Taliansky, S.I. Malyshenko, A.R. Mushegian, and O.A. Kondakova

Department of Virology, Moscow State University, Moscow 119899, USSR

At the moment of inoculation of a plant with a virus only a negligible number of cells become infected. The virus replicates in these initially-infected cells and then moves to the neighbouring healthy ones. A separate virus-specific function, namely the transport function (TF), is coded by the viral genome; "transport" proteins are encoded in the genomes of various plant viruses. In this paper three main questions will be discussed.

THREE TENTATIVE GROUPS OF TRANSPORT PROTEINS

Nucleotide sequence analysis of tobacco mosaic virus (TMV) mutants temperature sensitive (ts) in transport function (TF) revealed the presence of a single point mutation within the 30kd protein gene of TMV (Meshi *et al.*, 1987; Zimmern and Hunter, 1983). The more direct line of evidence for the transport role of TMV 30kd protein came from studies of R.N. Beachy and his colleagues of transgenic plants expressing a chimeric 30kd protein gene (Deom *et al.*, 1987).

Obvious homology has been revealed between the TMV 30kDa transport protein and the 29kDa transport protein encoded by tobacco rattle tobravirus (TRV) (Boccara *et al.*, 1986). Sequence similarity has been found also between the TMV 30kd protein and the putative transport proteins of several unrelated plant viruses, including caulimoviruses, como-, nepo- and potyviruses. Based on this sequence homology one can presume that the transport proteins of tobamo-, tobra-, caulimo-, and possibly como-, nepo- and potyviruses may form a separate transport protein group (Group I). The putative transport proteins of tricornaviruses (bromo-, cucumo-, and ilarviruses) have some sequence homology. Since these proteins are not homologous to the transport proteins of Group I, we propose to classify them as Group II transport proteins. There are good reasons to suggest that potex-, carla-, hordei- and furoviruses code for transport proteins which can be included in Group III. It is possible that differences in structure of virus-specific transport proteins reflect the existence of several distinct mechanisms of transport, resulting, however, in a similar effect: transfer of viral genetic material from cell to cell.

NONSPECIFICITY OF COMPLEMENTATION OF TRANSPORT FUNCTION

Virus-encoded transport function is one of the factors that controls virus host range; a plant can be resistant to a virus which cannot express TF in the plant. As a result the virus replicates only in the initially infected cells of the resistant plant. This type of resistance can be overcome if another virus that can normally spread in this plant complements the transport function of a transport-deficient virus. Similarly, systemic spread of TMV mutants, ts in TF, can also be complemented by a helper virus at restrictive temperature. Phloem-limited viruses which cannot move from cell to cell in parenchyma tissues alone acquire such ability in the presence of helper viruses. It is significant that the viruses of different "transport groups" can complement each other. For example, systemic spread of TMV ts-mutant *Ls1* can be complemented not only by tobamoviruses, but also by tobra-, como-, bromo- and potexviruses. Transport of red clover mottle comovirus (RCMV) in non-host plants is complemented by tobamo-, nepo- and cucumoviruses. Phloem-limited potato leaf roll luteovirus is transported into parenchyma cells in the presence of tobamo-, potex-, poty- and other viruses. Thus, one can conclude that the phenomenon of complementation of transport function is rather nonspecific (for review, see Atabekov and Talianski, 1990).

THE ROLE OF THE HOST IN TF EXPRESSION

It has been suggested that two genomes (that of a host and that of a virus) are involved in TF expression (Stussi-Garaud *et al.*, 1987; Tomenius *et al.*, 1987; Linstead *et al.*, 1988). We demonstrated that TMV mutants *Ls1* and Ni2519, ts in TF in tobacco, are temperature resistant in TF in another host, *Amaranthus caudatus*. Probably, some host factor(s) of this plant stabilized the transport protein of TMV ts mutants at the restrictive temperature. In the next series of experiments, we used *Ls1* to complement RCMV in *A. caudatus* plants at a nonpermissive temperature. It was expected that suppression of the temperature sensitivity of *Ls1* in *A. caudatus* would lead to its ability to promote RCMV transport in this plant. It was shown that RCMV can be systemically transported in *A. caudatus* in the presence of *Ls1* at 24°C or 33°C. Consequently the ts-defect in transport protein (30K protein) of TMV can be efficiently suppressed by a certain host, i.e., it is host-dependent.

cAMP plays an essential role in TF expression. Nicotinic acid, which decreases the level of cAMP in cells, blocks the systemic spread of TMV in leaf tissues. On the other hand, dibutyril cAMP overcomes the antiviral effect of nicotinic acid. These results imply that cAMP is involved in TMV TF expression. Therefore, TMV mutant *Ls1* (ts in TF) was used to examine the possibility of rescue of its TF by exogenous cAMP, papaverin and cholera toxin. Papaverin and cholera toxin are known to increase the intracellular level of cAMP. The transport of *Ls1*, which is blocked at the nonpermissive temperature in control plants can be rescued ("complemented") by exogenous cAMP, papaverin and cholera toxin. It should be noted that neither cAMP nor papaverin influenced the efficiency of *Ls1* replication in isolated protoplasts at nonpermissive temperatures.

REFERENCES

Atabekov J.G., and Taliansky M.E. (1990). *Advances in Virus Research* (In press).

Boccara, M., Hamilton, W.D.O., and Baulcombe, D.C. (1986). The organization and interviral homologies of genes at the 3' end of tobacco rattle virus RNA 1. *EMBO J.* **5**: 223-230.

Deom, C.M., Oliver, M.J., and Beachy, R.N. (1987). The 30-kilodalton gene product of tobacco mosaic virus potentiates virus movement. *Science* **237**: 389-394.

Linstead P.J., Hills G.J., Plaskitt K.A., Wilson I.G., Harker C.L., and Maule A.J. (1988). The subcellular location of the gene 1 product of cauliflower mosaic virus is consistent with a function associated with virus spread. *J. Gen. Virol.* **69**:1809.

Meshi, T., Watanabe, Y., Saito, T., Sugimoto, A., Maeda, T., and Okada, Y. (1987). Function of the 30 kd protein of tobacco mosaic virus: involvement in cell-to-cell movement and dispensability for replication. *EMBO J.* **6**: 2557-2563.

Stussi-Garaud C., Garaud J., Berna A., Godefroy-Colburn T. (1987). In situ location of an alfalfa mosaic virus non-structural protein in plant cell walls - correlation with virus transport. *J. Gen Virol.* **68**:1779-1784.

Tomenius, K., Clapham, D., and Meshi, T. (1987). Localization by immunogold cyto-chemistry of the virus-coded 30K protein in plasmodesmata of leaves infected with tobacco mosaic virus. *Virology* **160**: 363-371.

Zimmern, D., and Hunter, T., (1983). Point mutation in the 30K open reading frame of TMV implicated in temperature-sensitive assembly and local lesion spreading of mutant Ni 2519. *EMBO J.* **2**: 1893-1900.

DISCUSSION OF J. ATABEKOV'S PRESENTATION

Z. Xiong: You showed the rescue of the LS-1 transport mutant by comparison of the amount of virus accumulating in the leaf tissues. Were these tissues derived from the inoculated or the systemic leaves?

J. Atabekov: They were discs from the inoculated leaves.

Z. Xiong: I am asking because maybe cAMP is not rescuing the transport, but stimulating the virus multiplication.

J. Atabekov: There were controls. No influence of papaverine or the other agents were found in the virus replication in a TMV protoplast system.

SUMMARY AND CONCLUDING REMARKS BY THE FIRST SESSION CHAIRMAN, MILTON ZAITLIN

It is now an accepted fact that most plant viruses code for proteins that potentiate their movement from cell-to-cell in their hosts. This is an important phenomenon, not only because of its intrinsic interest in our understanding of virus movement, but because it has practical importance in explaining some forms of viral resistance; i.e., some plants are resistant to the disease that a virus might cause because the virus is unable to move from the initially-infected cells.

A number of studies employing immunogold tagging have shown that viral-coded movement proteins are associated with plasmodesmata in infected tissues. A paper by Beachy detailed his collaborative studies with W.J. Lucas involving tobacco plants transformed with the TMV 30K movement protein. Plants expressing the 30K protein have been injected with fluorescent dyes coupled to dextrans of differing molecular weights. The transformed plants allowed the movement of dye-dextran conjugates of 9400D, while in control plants the cut off values were between 700 and 800D (Wolf *et al.*, Science 246: 377-379, 1989). In TMV 30K-transformed plants, the age of the leaf influenced the degree of cell-to cell-movement; upper leaves were less able to permit the movement of the conjugates, and the virus did not move well in those leaves. While these findings do not explain either the effect of the movement proteins on the plasmodesmata which allows for the movement of such large particles as viruses or their nucleic acids, or account for the specificity of the movement of viral infections, they do indicate that the movement protein loosens the control of the plasmodesmata in permitting molecules to pass from one cell to its neighbor. Beachy also described studies on the capacity of the TMV 30K-transformed plants to affect the movement of other viruses. Except for potato leaf roll virus, the presence of the TMV 30K protein does not allow the cell-to-cell movement of viruses which do not normally move in tobacco. Furthermore, the effect on the movement of the leaf roll virus was only observed when the plants were inoculated by aphids, and not by mechanical inoculation.

The coat protein's role in long distance movement was also addressed. Dawson reported that when he substituted the coat protein gene of the orchid strain of TMV for the coat protein gene of the common strain, the orchid strain coat protein was able to encapsidate the chimeric RNA and form virions, but this virus would not move in tobacco, consistent with the behavior of the orchid strain in that plant. Both Beachy and Okada also stressed that the viral coat protein is important in the movement of virus in/out of the vascular system. Virus particles must be present for long distance movement to occur, but the observation of Dawson stated above shows that the coat protein could have an additional role in the process.

Not all viruses produce movement proteins — in particular the phloem-limited luteoviruses do not, as pointed out by Harrison. There is some evidence however, that viruses which do code for movement proteins can help other viruses to move. Atabekov has documented a number of instances in which this is possible by showing that systemic infection by a helper virus can allow a dependent virus, absorbed through the vascular system, to move into mesophyll cells. The relationship of this phenomenon to the normal mesophyll cell movement, and its dependence on a movement protein have not been established. Atabekov has demonstrated further that the activity of the movement protein can be host specific, suggesting that a specific host factor or component is involved; i.e., a TMV strain which has a temperature sensitive defective movement protein rendering it unable to move in tobacco at a restrictive temperature can move in *Amaranthus caudatus* at that temperature. He demonstrated further that nicotinic acid blocks TMV movement and that cyclic AMP, or either papaverine or choleratoxin, both of which increase cyclic AMP, can overcome this inhibition.

Okada addressed the issue of the functional domains of the TMV 30K movement protein. His laboratory found that if they deleted the 31 C-terminal amino acids, the protein was still functional in promoting virus movement. Dawson, who has generated infectious transcripts from full length clones of TMV, has made modifications in the TMV 30K protein gene. He has moved the gene to different positions in the genome, and has inserted extra copies of the 30K protein. Although some of his mutants reduced the production of the 30K protein in the host, only those with less than 10% of wild-type protein showed a reduced capacity to affect virus movement. The position of the 30K gene with respect to the 3' end of the virus RNA affected the level of protein produced; more protein was synthesized the closer the gene was to the 3' end of the RNA.

A substantial number of genes are induced in plants which show a hypersensitive response when infected by viruses. Among these are genes which code for a number of "pathogenesis-related" or PR proteins. Over 20 have been detected in TMV-infected tobacco. Their role in resistance, if any, remains to be elucidated. Bol and his colleagues have transformed plants with genes encoding either one of two PR proteins (PR-1 and PR-S) or a glycine-rich protein, also known to be induced by TMV infection. Proteins coded by these genes are constitutively expressed in tobacco plants which were tested for their response to both virus infection and to insect attack by the tobacco budworm. They concluded that the expression of any of these proteins individually in the plant is not responsible for resistance to virus infection, nor are they a possible defence of tobacco to insect attack. One gene, PR-S was also inserted into the plant in an "antisense" orientation. Production of PR proteins R and S was reduced in such plants up to 80% but their response to the virus in forming necrotic lesions was unaffected.

Transgenic Host Response to Gene VI of Two Caulimoviruses

Karen-Beth Goldberg, Jennifer Kiernan, J.E. Schoelz and R. J. Shepherd

Department of Plant Pathology, University of Kentucky,
Lexington, KY 40546, USA

Three separate laboratories have reported that transformation of *Nicotiana tabacum* with gene VI of cauliflower mosaic virus (CaMV) causes the induction of a chlorotic mottling syndrome similar to the chlorosis induced by virus infections (Goldberg *et al.*, 1987; Baughman *et al.*, 1988; Takahashi *et al.*, 1989). A relationship was established by these investigators between the level of expression of the gene VI protein and the presence of the disease syndrome. These observations suggested that P62, the gene VI protein, perturbs plants to cause disease and that perhaps gene VI is a chlorosis gene that accounts for the chlorotic effects commonly associated with the diseases induced by caulimoviruses. In the present study we have transformed additional plants, both systemically susceptible and non-hosts species, to further investigate this disease induction phenomenon.

DISEASE RESPONSES ASSOCIATED WITH GENE VI OF CAULIMOVIRUSES

We have carried out experiments with both CaMV and figwort mosaic virus (FMV) that suggest gene VI is a major element in disease development. In addition, its role as a major host range determinant is well established (Schoelz *et al.*, 1986; Schoelz and Shepherd, 1988). Moreover, region VI determines the nature of initial host reaction, whether it is hypersensitive or compatible (Schoelz *et al.*, 1986) suggesting that gene VI may be the main viral determinant that mediates the virus-host interaction.

Mutagenesis

The first suggestion that gene VI of CaMV was an important factor in disease induction came from a series of mutagenesis experiments with a particularly severe strain of this virus. Small in-phase insertions made into the cloned genome of the Cabb B-Davis strain of CaMV, were assayed on turnip (*Brassica campestris*). Two of the three mutants containing 12 base pair insertions in gene VI were infectious on turnip but caused much less severe disease (Daubert *et al.*, 1983). One mutant caused only mild chlorotic mottling without stunting or necrosis. In contrast, the wild type parent clone induced severe stunting, mottling and necrosis (Daubert *et al.*, 1983). Although these observations suggested that gene VI may be a particularly important factor in disease, these mutants were less productive than wild type virus. Consequently, it was not clear whether gene VI had a direct role in disease development.

More recently mutants of CaMV have been described that affect symptoms as well as the ability of CaMV to develop systemically in *Datura stramonium* (Daubert, 1988).

Hybrid genomes

Several strains of CaMV with unique properties have been recombined *in vitro* (as the cloned segments) to yield hybrid genomes. The phenotypic characters of these hybrids in comparison to the parental wild-types have been used to map some host range and disease determinants.

Newly constructed hybrid strains have been analyzed with respect to disease production in an effort to associate particular disease syndromes with particular portions of the CaMV genome. Much of our early work suggested that if one started with a productive but symptomless strain (e.g., the H7 recombinant is very productive but masked in infected *Brassica campestris* var. "Just Right"), stepwise segmental exchanges could be associated with particular disease determinants (Daubert *et al.*, 1984; Schoelz *et al.*, 1986). Experiments of this sort with gene VI of CaMV strain D4 appeared to confer an avirulent phenotype to hybrids made with it, whereas if gene VI came from a virulent strain the resulting hybrid genomes would be virulent. Minor variations in disease manifestations were attributed to the influence of other genes on virus productivity, invasiveness, etc., that affected the copy number and expression of gene VI. To a great extent similar observations of others have been compatible with this view (Stratford & Covey, 1989). Now, however, as demonstrated with more recently constructed hybrid genomes (E. J. Anderson and J. E. Schoelz, unpublished data) we are less convinced of this interpretation of disease induction. We now suspect that toxic versions of other genes exist that can cause chlorosis and that gene VI may not be the only factor involved in the production of conventional disease symptoms.

DISEASE RESPONSES OF GENE VI TRANSFORMED PLANTS

For genetic transformations of plants, gene VI of CaMV or FMV was cloned into the binary transformation vectors pGA472 (An *et al.*, 1985) and/or pKYLX-7 (Schardl *et al.*, 1987). A conventional *Agrobacterium*-mediated leaf disc transformation procedure was used (Horsch *et al.*, 1985).

Expression of Gene VI in *N. tabacum*

Gene VI of CaMV (pJS62 and pJS65: Fig. 1) and FMV (pKB39: Fig. 1) were used to transform tobacco cv. Burley 21. The regenerated plants were uniformly kanamycin resistant. Symptom development in these transgenic plants could be correlated to gene VI accumulation. Plants with greater accumulations of gene VI protein had more severe symptoms which included stunting, a chlorotic mottle, narrowed leaves and/or delayed flowering (Fig. 2).

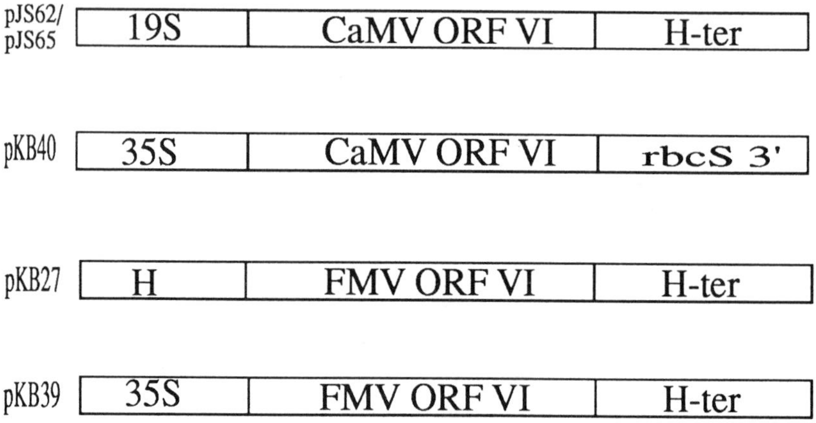

Fig. 1 The gene VI constructs used for transformation of the three solanaceous species. pJS62 and pJS65 represent gene VI of CaMV strains CM1841 and D4, respectively, with a 19S CaMV promoter and homologous transcriptional terminator. pKB27 consists of FMV gene VI with its homologous (H) promoter and terminator (H-ter). All three sources of gene VI were transferred into pGA472 (An *et al.*, 1985), a plant transformation vector. pKB40 and pKB39 represent CaMV CM1841 and FMV, repectively, with a 35S CaMV promoter and either a rubisco 3' or a homologous transcriptional terminator (H-ter). Gene VI of both viruses were transferred into pKYLX-7, a plant transformation vector (Schardl *et al.*, 1987).

A few plants developed generalized chlorosis, particularly with gene VI of FMV. In addition, the chlorophyll level of the pKB39 Burley 21 plants was reduced by approximately 60% as compared to healthy Burley 21 leaves. Over half of the F_1 progeny of these plants showed symptoms similar to the initial transformants.

Burley 21 is not a systemic host of any strain of CaMV or FMV. Moreover, no local symptoms are usually observed on inoculated leaves with FMV and most strains of CaMV. However, one strain of CaMV, NRt-17, does induce chlorotic local lesions on inoculated leaves of tobacco.

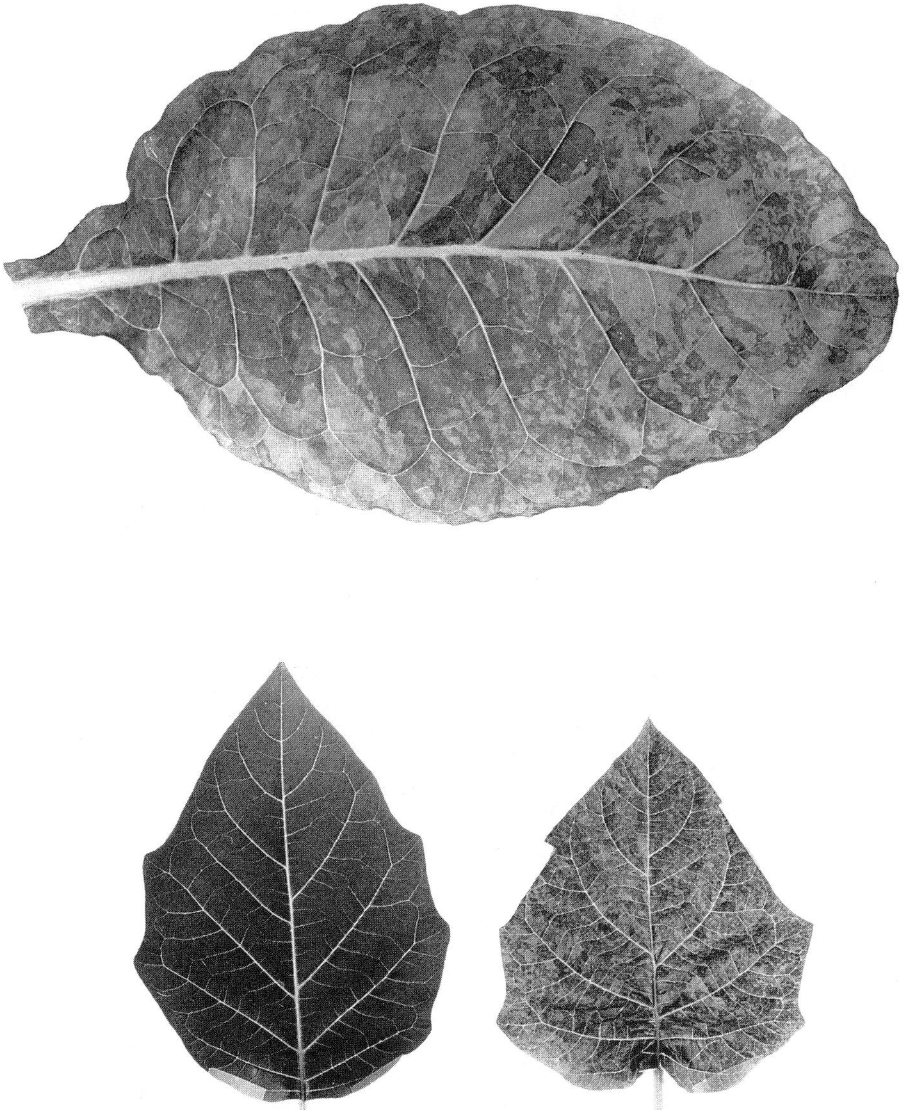

Fig. 2 Typical gene VI-transformed plants showing the chlorotic mottling syndrome associated with expression of gene VI. The leaf at the top is Burley 21 tobacco after transformation with gene VI of CaMV (from Shepherd *et al*., 1988, by permission) The lower leaves are from healthy *Datura innoxia (*left) and from a plant transformed with gene VI of CaMV (right).

Expression of Gene VI in *Datura innoxia*

When CaMV (strains D4 or CM1841) gene VI was used to transform *D. innoxia*, the plants that expressed significant levels of gene VI protein exhibited chlorotic mottling. Transformants of pJS62 and pJS65 expressing the highest levels of P62 (Fig. 3) showed the most severe mottling. In addition, one pKB40 transformant (Fig. 1), under control of the CaMV 35S promoter, developed a severe necrosis and stunting.

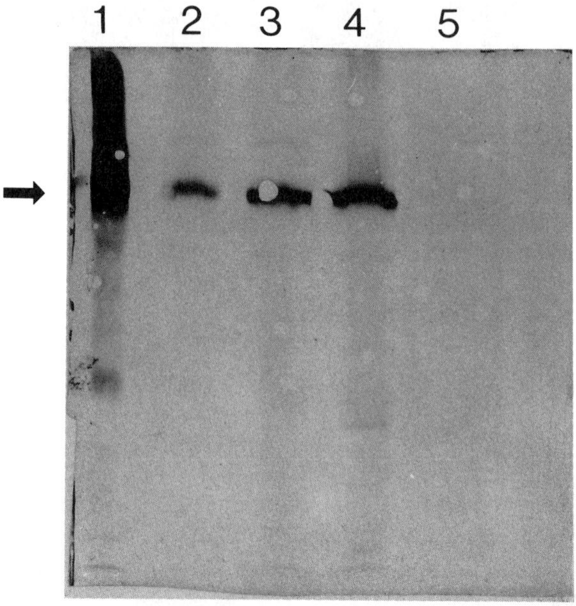

Fig. 3 Western blot analysis of gene VI protein accumulation in *Datura innoxia* transformed with pJS62 and pKB40 (gene VI of CaMV CM1841 under the transcriptional control of the 19S or 35S promoter, respectively). Each lane contained 40 µl of extract (0.5 g leaf/1.5 ml Laemmli buffer) unless otherwise indicated. Lane 1, 2.0 µl CaMV CM1841 infected turnip; Lanes 2 and 3, plants #3 and #10, respectively, of *D. innoxia* transformed with pKB40; lane 4, plant #8 of *D. innoxia* transformed with pJS62; lane 5, non-transformed *D. innoxia*. The arrow on the left margin indicates the position of P62, the protein of gene VI.

The prominent effects associated with transformations with gene VI of CaMV were in marked contrast to the lack of effects when this species was transformed with gene VI of FMV (pKB27 and pKB39; Fig. 1). The latter transformants never developed observable chlorosis or stunting even though they accumulated gene VI-specific RNA and protein to levels similar to those transformed with gene VI of CaMV. The key difference between the two viruses is that FMV can systemically infect *D. innoxia* whereas CaMV does not. Hence there is an inverse relationship between systemic susceptibility and the induction of the chlorotic mottling syndrome from transformation with gene VI in *D. innoxia.*

The Effect of Gene VI in *Nicotiana edwardsonii*

The transformation of *N. edwardsonii* with gene VI of CM1841 (pJS62 and pKB40; Fig. 1) resulted in plants which displayed either a mild chlorosis or mild chlorotic vein clearing which was associated with low levels of P62 accumulation. CaMV CM1841 cannot systemically infect *N. edwardsonii.*

When FMV (pKB27) or CaMV D4 (pJS65; Fig. 1) gene VI was used to transform *N. edwardsonii* the plants showed no visible effects though they accumulated low levels of P62. Again, this suggests an association between development of chlorosis and lack of systemic susceptibility.

A few transformed plants of *N. bigelovii* were obtained using pJS65, containing gene VI of D4 CaMV. This species is systemically susceptible to D4 CaMV. Again no detectable effects were observed in plants which accumulated significant levels of P62.

From our experiments the induction of the chlorotic mottling syndrome or generalized chlorosis appears to be a non-host reaction to gene VI expression. This raises doubt about the biological significance of the phenomenon. However, the level of gene VI protein in our transgenic plants was never more than one-tenth the level produced in virus-infected plants. It may be that non-hosts are more sensitive to being perturbed by gene VI protein than are systemically susceptible hosts. Perhaps if the level of expression could be greatly elevated in transformed hosts, these might also respond with chlorosis. However, in our trials using the 35S promoter of CaMV to drive expression of gene VI we were able to achieve only slightly higher levels of expression than with the homologous promoters. Consequently, this issue has not been resolved by our experiments.

REFERENCES

An, G., Watson, B. D., Stachel, S., Gordon, M. P., and Nester, E. W. 1985. New cloning vehicles for transformation of higher plants. *EMBO J.* **4**:277-284.

Baughman, G. A., Jacobs, J. D., and Howell, S. H. 1988. Cauliflower mosaic virus gene VI produces a symptomatic phenotype in transgenic tobacco plants. *Proc. Natl. Acad. Sci. USA* **85**:733-737.

Daubert, S. 1988. Sequence determinants of symptoms in the genomes of viruses, viroids, and satellites. *Molec. Plant-Microbe Interactions* **1**:317-325.

Daubert, S. D., Shepherd, R. J., and Gardner, R. 1983. Insertional mutagenesis of the cauliflower mosaic virus genome. *Gene* **25**:201-208.

Daubert, S. D., Schoelz, J. E., Li, D., and Shepherd, R. J. 1984. Expression of disease symptoms in cauliflower mosaic virus genomic hybrids. *J. Molec. Appl. Gen.* **2**:537-547.

Goldberg, K.-B., Young, M., Schoelz, J. E., Kiernan, J. M., and Shepherd, R. J. 1987. Single gene of CaMV induces disease. *Phytopathology* **77**:1704.

Gowda, S., Goldberg, K.-B., Kiernan, J., Schoelz, J., Young, M., Richins, R., and Shepherd, R. J. 1989. Pathogenesis and host range determination by caulimoviruses. In B. Staskawicz, P. Ahlquist, and O. Yoder (eds.): "Molecular Biology of Plant-Pathogen Interactions", pp. 229-243. Alan R. Liss, Inc., New York.

Horsch, R. B., Fry, J. E., Hoffmann, N. L., Eichholtz, D., Rogers, S. G., and Fraley, R. T. 1985. A simple and general method for transferring genes into plants. *Science* **227**:1229-1231.

Schardl, C. L., Byrd, A. D., Benzion, G., Altschuler, M. A., Hildebrand, D. F., and Hunt, A. G. 1987. Design and construction of a versatile system for the expression of foreign genes in plants. *Gene* **61**:1-11.

Schoelz, J. E., Shepherd, R. J., and Daubert, S. D. 1986. Region VI of cauliflower mosaic virus encodes a host range determinant. *Mol. Cell. Biol.* **6**:2632-2637.

Schoelz, J. E., Shepherd, R. J., and Daubert, S. D. 1987. Host response to cauliflower mosaic virus (CaMV) in solanaceous plants is determined by a 496 bp DNA sequence within gene VI. In C. J. Arntzen and C. A. Ryan (eds.): "Molecular Stategies for Crop Protection", pp. 253-265. Alan R. Liss, Inc., New York.

Schoelz, J. E. and Shepherd, R. J. 1988. Host range control of cauliflower mosaic virus. *Virology* **162**:30-37.

Shepherd, R. J., Goldberg, K., Kiernan, J., Gowda, S., Schoelz, J., Young, M., Richins, R., (1988). Genomic changes during host adaptation to caulimoviruses, pp. 131-138. In Physiology and Biochemistry of Plant - Microbial Interactions. Amer. Soc. of Plant Physiologists, Rockville, MD.

Stratford, R. and Covey, S. N. 1989. Segregation of cauliflower mosaic virus symptom genetic determinants. *Virology* **172**:451-459.

Takahashi, H., Shimamoto, K., and Ehara, Y. 1989. Cauliflower mosaic virus gene VI causes growth suppression, development of necrotic spots and expression of defense related genes in transgenic tobacco plants. *Mol. Gen. Genet.* **216**:188-194.

DISCUSSION OF K. GOLDBERG'S PRESENTATION

R. Hull: I wish to report on some observations made by Simon Covey and colleagues at the John Innes Institute on the interactions of CaMV with various brassica species. The Cabb BJI isolate of CaMV gives a severe infection of turnip cv Just Right leading to general chlorosis of the leaves and severe stunting of the plant. In swede (rutabaga) or rape the symptoms are very mild with chlorotic vein banding and little stunting. In kohlrabi there are no apparent symptoms even though virus can be isolated from the plants.

The replication of CaMV DNA has two phases, that in the nucleus where RNA is transcribed from the viral DNA and that in the cytoplasm where the RNA is reverse-transcribed to give viral DNA. The 2-D electrophoresis technique, in which DNA is electrophoresed in the first dimension in a neutral buffer and in the second dimension under denaturing alkali conditions, was used to identify covalently closed circular viral DNA from the nuclear phase of replication and the various open circular and linear DNA forms which arise in the cytoplasmic phase of replication. 2-D gel electrophoresis of DNA from infected turnip plants shows that in this host the cytoplasmic forms predominate. In rape there is a higher proportion of the nuclear forms of DNA to the cytoplasmic forms and in kohlrabi most of the viral DNA is in the nuclear form. These are reflected in the levels of transcription with much viral RNA being found in turnips and scarcely detectable amounts in kohlrabi.

Even in the permissive host turnip there was tissue dependence of the proportions of nuclear and cytoplasmic forms of DNA. In leaves the cytoplasmic forms predominated, in the stems the proportions changed in favour of the nuclear forms and in the roots it was the nuclear form which predominated.

These observations show that even in full systemic hosts there are considerable interactions between the viral and plant genomes.

U. Melcher: Do the protein levels in the transgenic plants correlate with the RNA levels and could symptoms be due to RNA rather than protein?

K. Goldberg: When there are high levels of protein there are high levels of RNA but I don't have any specific data which can answer the question.

T. Hohn: How do the supercoiled forms of CaMV DNA accumulate?

R. Hull: Even in the least permissive host there is some of the cytoplasmic form of the virus which could feed back to the nucleus. It is a balance between the nuclear and cytoplasmic forms which in kohlrabi is tilted greatly to the nuclear forms.

C. Collmer: Do plants which normally do not support systemic spread of CaMV support spread when they are transgenic for gene VI?

K. Goldberg: In the limited experiments we have done no evidence has been found for the systemic spread of CaMV in gene VI transgenic 'non-hosts'.

R. Gilbertson: Have you made any exchanges between CaMV and FMV gene VI?

J. Schoelz: No.

R. Gilbertson: How related are CaMV and FMV and why have they been named different viruses?

S. Gowda: I will discuss this in my talk.

T. Hohn: Is there stability of expression in the progeny of gene VI transgenic plants or are there some progeny which, when assessed by southern blotting, are not expressing any more, due to say methylation?

K. Goldberg: Progeny express gene VI and no evidence has been noticed for lack of expression due to methylation.

R. Gilbertson: What was the segregation of gene VI in transgenic plants?

K. Goldberg: The inheritance has not been studied in detail. In *N. tabacum* cv Burley 21 we often obtained a 3:1 ratio which would suggest a single dominant gene. However, for plants which were very severely affected it appeared that the gene dosage was accumulating and the ratios became more skewed in the next generation.

The Use of 35S RNA as Either Messenger or Replicative Intermediate Might Control the Cauliflower Mosaic Virus Replication Cycle

T.Hohn, J-M.Bonneville, J. Fütterer, K.Gordon, J.Jiricny, S.Karlsson, H.Sanfaon, M.Schultze and M. de Tapia

Friedrich-Miescher-Institut, PO Box 2543, CH4002, Basel, Switzerland

The plant pararetrovirus cauliflower mosaic virus (CaMV) contains 8000 bp of open circular DNA, which become supercoiled upon entering the nucleus (reviewed by Bonneville *et al.*, 1988). Two major transcripts have been characterized (Fig. 1): the 19S RNA, which encodes the viral inclusion body matrix protein and the terminally redundant 35S RNA, which is the template for the viral reverse transcriptase and probably the mRNA for most of the viral proteins, especially systemic spreading factor, insect transmission factor and capsid protein. In addition a third RNA (22S RNA) of low abundance, covering ORF V, might exist (Plant *et al.*, 1985). The 35S RNA is a very poor substrate for translation *in vitro* (Gordon *et al.*, 1988), probably because of its 600 nt long leader containing several small ORFs and due to its polycistronic nature. We therefore studied its translation in transfected plant protoplasts with reporter genes attached to various viral ORFs. Our results indicate that a transactivator encoded by the 19S operon is required for efficient expression of most downstream ORFs. Availability of this transactivator together with some cellular factors may determine the use of the 35S RNA as either template for reverse transcription or as mRNA for structural proteins and thereby divide the CaMV replication cycle into an early and a late phase.

HYPOTHETICAL DIVISION OF THE CAMV REPLICATION CYCLE INTO AN EARLY AND A LATE PHASE

According to our proposition (Fig. 2), the early phase of CaMV replication starts with the establishment of the first copy of the CaMV minichromosome in the nucleus (Menissier *et al.*, 1982) and the first copy of 35S RNA in the cytoplasm (Guilley *et al.*, 1982). At this stage 35S RNA behaves as expected for a polycistronic RNA containing upstream AUGs and from the *in vitro* translation results (Gordon *et al.*, 1988), *i.e.* it is not translated. It is, however, reverse transcribed to yield open circular DNA (Pfeiffer & Hohn, 1983). Due to a deficiency of coat protein neither 35S RNA nor DNA is packaged into mature virus particles; instead, the open circular DNA is retransported into the nucleus and converted to supercoiled DNA (minichromosome). This transcription and reverse transcription-cycle continues and more and more minichromosomes are accumulated. Theoretically, only one viral gene product is required in this phase, reverse transcriptase, and accordingly we have to postulate that its gene, *i.e.* ORF V, is transcribed and translated during the early phase.

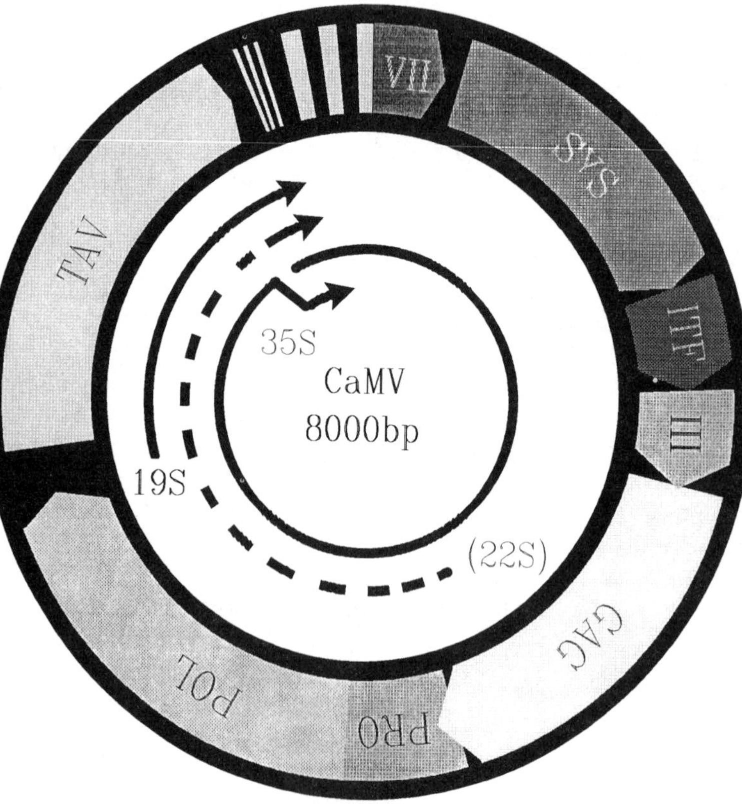

Fig. 1 Map of CaMV. The open reading frames (ORFs) and their functions are shown clockwise starting with O hours in the outer circle: VII (unknown function), I = SYS (systemic spreading function), II = ITF (insect transmission factor), III (unknown function, part of capsid?), IV = GAG (a polyprotein including capsid protein[s]), V = POL (polymerase, a polyprotein consisting of PROtease and reverse transcriptase/RNAse H), VI = TAV (transactivator and inclusion body matrix protein), six small ORFs (A,B,C,D,E,F, positionally conserved in most CaMV strains). The arrows inside the circle show the transcripts named after their sedimentation constant. The 22S RNA is still hypothetical.

Simultaneously with 35S RNA, 19S RNA is also produced. 19S RNA codes for the main component of the inclusion body, within which virus particles accumulate during late stages of infection (Covey *et al.*, 1981). The 19S promoter is much weaker than the 35S one (Lawton *et al.*, 1987), and it therefore takes some time until larger amounts of 19S RNA and ORF VI protein accumulate. When enough ORF VI protein is produced, it will promote translation of the 35S RNA (see below), thereby switching the replication cycle from early to late. One of the 35S translation products is capsid protein, which now stably packages the CaMV genome either as RNA, as open circular DNA or as some other type of replication complex. As a consequence, DNA copies cease to enter the nucleus of the original cell but they spread as virus particles from cell to cell or from plant to plant. The proteins necessary for these processes, systemic spreading and aphid transmission factors, also become available during the late phase.

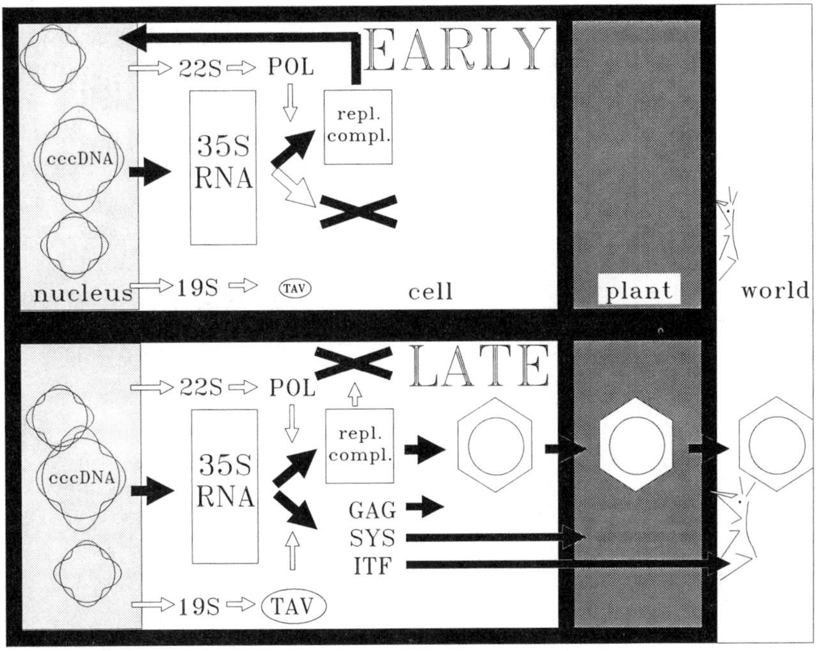

Fig. 2 Early and late phase of CaMV replication as proposed. Early: the constitutive steps are shown accumulating supercoiled DNA in the nucleus; translation of ORFs from the 35S RNA is inhibited (X) but translation from 22S and 19S RNAs occurs to yield polymerase (POL) and transactivator (TAV). Late: When large amounts of transactivator (TAV) protein have been accumulated, translation from 35S RNA occurs. The capsid protein (GAG), which is now produced, assembles and packages the replication complexes; accordingly, further accumulation of virus DNA in the nucleus (supercoiled DNA) is blocked (X). Systemic spreading (SYS) and insect transmission (ITF) proteins function to spread the virus within the plant and within the field.

The complex transactivation mechanism could also provide a means of control at the cell-specific level and could be used to differentially express the various ORFs. For instance, it might be appropriate to accumulate supercoiled DNA in the nuclei of dividing cells but virus particles in the cytoplasm of elongating cells; it might be convenient to produce virus particles and insect transmission factor predominantly in leaves, the feeding ground of the transmitting insects; it might be desirable to produce in the originally infected leaf predominantly systemic spreading factor to allow fast transport of virus to the very young tissue close to the meristematic apex (Maule, 1985; Maule *et al.*, 1989), etc.

RECENT EVIDENCE FOR THE MODEL

Expression of Reverse Transcriptase

Since ORF IV overlaps ORF V by 12 codons, it was originally assumed that ORF V is translated as an ORF IV/V fusion protein much like the gag/pol fusion protein of retroviruses (Jacks & Varmus, 1985). However, no such fusion protein could be detected either in infected cells (Pietrzak & Hohn, 1987, Kirchherr *et al.*, 1988) or *in vitro* (Gordon *et al.*, 1988). Further, Penswick *et al.*, (1988) described a viable CaMV mutant isolate in which ORF IV and V do not overlap and Schultze *et al.*, (1989) constructed a viable and stable virus in which ORF IV and V were separated by an oligonucleotide containing stop codons in all three reading phases. They also showed that ORF V start codon mutants are either lethal or give rise to true and second-site revertants to restore or create start codons. Translation from CaMV ORF V (pol) therefore appears to be independent from that of the overlapping, upstream ORF IV (gag). This correlates with the presence of the start codon (in fact two) at the beginning of ORF V, which has no counterpart in the retroviral cases.

Recent experiments suggest that CaMV ORF V is also transcribed separately. Plant *et al.*, (1985) reported evidence for a third CaMV RNA (22S RNA) covering ORF V (and ORF VI?) (Fig. 1) and some of us (M.S., J.J.& T.H., in prep) have evidence for a weak CaMV promoter in front of ORF V. This promoter directs α-glucuronidase (GUS) reporter activity in transfected *Orychophragmus violaceus* protoplasts. It would be located within a stretch of more than 47 and less than 343 basepairs upstream of ORF V (Fig. 3). Its strength, 1% the strength of the 35 S promoter, would be low, but certainly sufficient for an enzyme gene. Thus independent transcription and translation of ORF V is possible allowing reverse transcriptase production before viral capsids can be formed, a requirement of our model.

A viral factor involved in translation from 35S RNA

To test conditions for translation of ORFs located on 35S RNA, plasmids were constructed to contain original CaMV sequences including promoter, leader and an increasing number of upstream ORFs. The coding region of either the chloramphenicol acetyltransferase (CAT) or GUS was fused to the start codon of the last ORF, followed by the CaMV polyadenylator (Fig. 4). By these means functional translational fusions to ORFs VII, I, II, III, IV and V were obtained. For each construct a positive control was produced having leader and

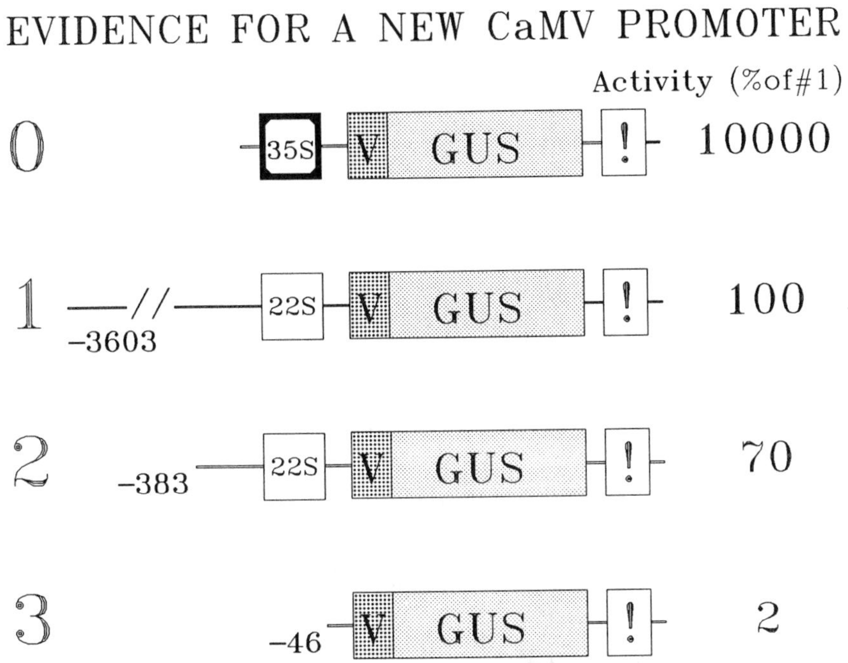

Fig. 3 Evidence for a new CaMV promoter. O) The α-glucuronidase (GUS) coding region was fused to the first codons of ORF V and put under 35S promoter and CaMV polyadenylator (!) control. 1,2,3) Similar constructs with the 35S promoter omitted but with 3603, 383 and 46 basepairs, repectively, of original upstream sequence retained. GUS activities are given in comparison to the value obtained with construct 1, which was set to 100%.

upstream ORFs removed. All plasmids were transfected into *O. violaceus* protoplasts and the reporter activities measured upon 16 hours. Only ORF VII- and ORF II- fusions gave rise to significant reporter activity, which amounted to 30 and 15% that of the respective monocistronic control (Fig. 4). In a second experiment all samples were cotransfected together with a transactivator plasmid consisting of CaMV ORF VI with its 19S promoter exchanged for the stronger 35S one (Bonneville *et al.*, 1989). All reporter expression levels were raised considerably, in some cases to 100% (Fig. 4). Transactivation of ORF I expression by the ORF VI product was also observed for the related figwort mosaic virus (FMV; Gowda *et al.*, 1989 and our own unpublished observation). Experiments showed that transactivation by the ORF VI product (hence called transactivator) occurs at the level of translation. Experiments by Bonneville *et al.*, (1989) on ORF I expression also showed that transactivation was much less pronounced when ORF VI was under its original 19S promoter control. This indicates that certain minimal amounts of transactivator are required for optimal performance, again in accordance with the model.

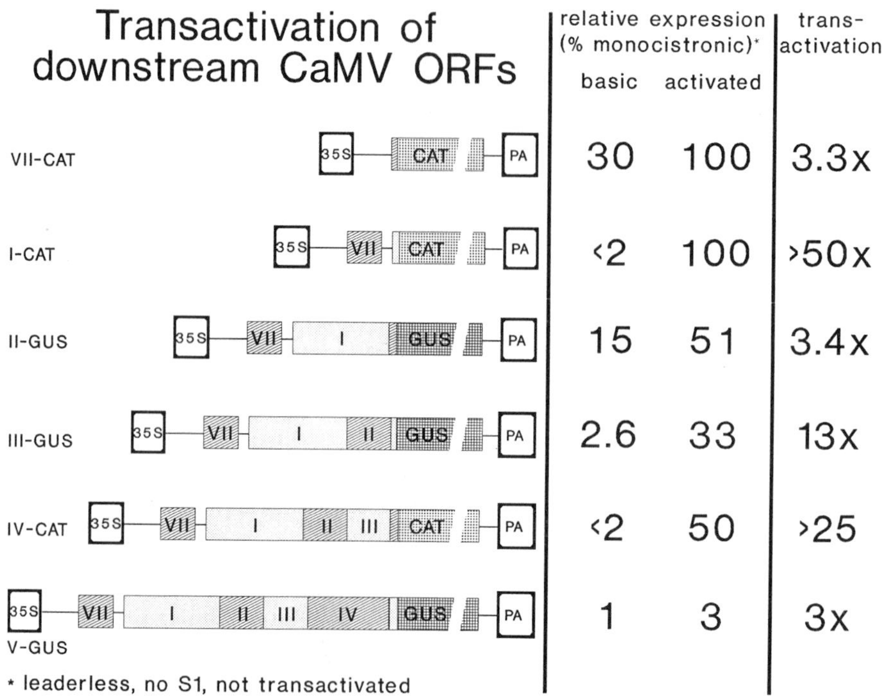

Transactivation of downstream CaMV ORFs
relative expression (% monocistronic)*
basic
activated
trans-activation
VII-CAT
35S
CAT
PA
30
100
3.3x
I-CAT
VII
<2
100
>50x
II-GUS
I
GUS
15
51
3.4x
III-GUS
II
2.6
33
13x
IV-CAT
III
<2
50
>25
V-GUS
IV
1
3
3x
* leaderless, no S1, not transactivated

Fig. 4 Transactivation of translation of downstream CaMV ORFs. Original CaMV sequences from the 35S promoter to various points within the CaMV genome were fused to either the chloramphenicol acetyltransferase (CAT) or α-glucuronidase (GUS) reporter coding region, and the CaMV polyadenylator sequence (PA) was attached. Fusions were choosen to be translational, attaching the reporter ORFs to a few N-terminal amino acids of each ORF. For each fusion a monocistronic derivative was produced as positive control (not shown). Plasmids were transfected into *O. violaceus* protoplasts, soluble cell extracts were prepared after overnight incubation and reporter activities determined. Results are expressed as % of the respective monocistronic control. First column, plasmids transfected alone. Second column, plasmids cotransfected with transactivator plasmid consisting of ORF VI under 35S promoter and polyadenylator control. Third column, ratios between basic and transactivated expression levels.

Are cellular factors involved in translation from 35S RNA?

The constructs containing the ORF VII-CAT fusions mentioned above, with their relatively high expression even in the absence of transactivation, were tested in different types of plant protoplasts (Fütterer *et al.*, 1989). Expression of CAT positioned downstream from the leader was consistently lower in Solanaceae compared to expression in *O. violaceus*, as described above, and other Cruciferae, such as *Brassica* and *Sinapis*. Using related constructs also in *Daucus carota*, only very low levels of reporter activity were observed

(Baughman & Howell, 1988). The different behavior of Cruciferae and other plants indicates that cellular factors are also involved in regulation of ORF VII expression. The relatively high expression level in Cruciferae is astonishing in view of the 9 AUG codons in the leader and in view of the leader symmetry (Fütterer *et al.*, 1988). To study this question further a large series of derivatives with various deletions within the leader were constructed and assayed. Results revealed that the leader consists of a mosaic of inhibiting (I1, I2) and stimulating (S1, S2, S3) sequences (JF, HS, J-MB, KG and TH, in prep. and Fig. 5). Not unexpectedly, the inhibiting sequences comprise the sORFs. Stimulator S1, comprising the first 60 leader nucleotides, activates expression

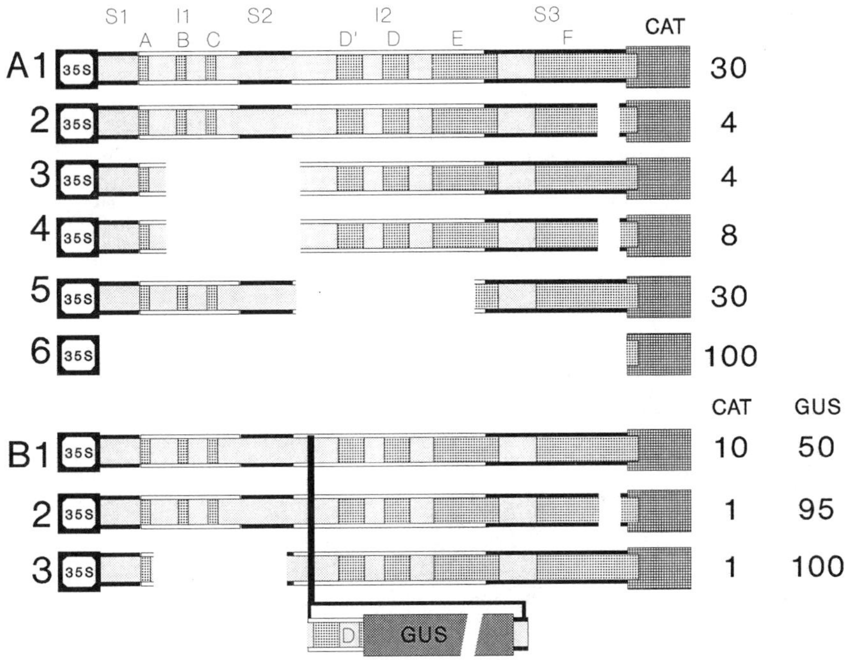

Fig. 5 Regulative elements within the CaMV leader sequence. A.) Constructs consist of the 35S promoter, the complete leader the CAT reporter ORF and the polyadenylator (1). Positions of the sORFs A,B,C,D',D,E and F are shown and stimulatory (S1,S2,S3) and inhibitory regions (I1,I2) are defined. Portions (2 to 5) and all of the leader (6) are deleted in the following constructs, as indicated by blank regions. B.) 1 to 3. Constructs as A 1 to 3, however the GUS ORF and flanking regions were inserted into the I2 region, as indicated. The constructs were transfected into *O. violaceus* protoplasts and CAT and GUS activities measured and compared.

two-fold in whatever type of protoplasts tested and in presence or absence of the rest of the leader. Sequences S2 and S3, in contrast, function only if present together and only in certain cell cultures, predominantly derived from host plants. They stimulate expression 10 - 20-fold and act in supressing inhibition by the sORFs in the I2 region of the leader. In the context of the complete leader, sequences within I2 can be deleted or added without significant effect on expression. In fact a second reporter gene can be introduced into I2 leading to a dicistronic mRNA producing still considerable amounts of the second (CAT) reporter activity in addition to the first one (GUS; Fig. 5). To explain these properties of the leader, a ribosome shunt model is proposed (JF, HS, J-MB, KG and TH in prep.), according to which ribosomes scan normally along the leader of 35S RNA until region S2 is reached. From this region a certain fraction proceed directly to S3 and initiate at ORF VII. Due to this shunt the inhibitory region I2 is skipped. We think that cellular factors are involved in the shunt, since it is not equally efficient in all types of plants. This mechanism allows ORF VII and perhaps ORF II to be translated at a detectable level in absence of viral transactivator. The biological purpose of such a non-transactivated translation level is not clear at this moment, however, the shunt mechanism observed in this case might basically be used in all translations from 35S RNA, with the transactivator enhancing the process. A more general idea on ribosome shunt in CaMV was published earlier by Hull (1984).

Evidence for tissue specific regulation of reverse transcription and translation

Rapidly dividing callus tissue can be derived from leaves either via protoplasts or from leaf-discs placed on nutrient agar. The leaf mesophyll cells apparently de-differentiate and resume DNA replication and cell division. In view of this de-differentiation it is interesting that in callus tissue derived from CaMV infected leaves, newly synthesized protein could not be detected and ratios of viral DNA types change dramatically in favour of supercoiled DNA and at the cost of open circular forms (Paszkowski *et al.*; 1983, Rollo & Covey, 1985). Smaller satellite supercoils accumulate too. It seems that in this tissue virus replication is converted back to the early phase, not allowing capsid protein production. Due to the focus of DNA replication to the single cell, selective pressure to maintain the complete genome was abolished, giving rise to the small variants. As an additional effect the small variants might predominantly retain the reverse transcriptase gene and replication signals and by outcompeting the complete genome shift the replication mode even more to the early phase.

Recently Covey *et al.*, (1990) analyzed CaMV DNA populations in various tissues of infected plants. Results revealed that systemically infected leaves produce relatively small amounts of supercoiled DNA, but large amounts of open circular DNA, stems produce intermediate amounts of both types and roots produce mainly supercoiled DNA. They again observed predominantly full and small size supercoils in callus tissue. The authors also showed that even in leaves of host plants supercoiled DNA predominates if these allow only weak systemic symptoms. These results could well be interpreted in terms of our model with tissue from roots of "good" host plants and tissue from all parts of poor host plants being blocked or at least inhibited in the late viral replication phase. These systems might therefore become valuable tools to study more of the molecular mechanisms of control of virus replication and expression.

TEMPORAL CONTROL IN OTHER VIRUSES; COMPARISON WITH HIV-1 TRANSACTIVATION

The best studied cases of differential gene expression in viruses involve DNA tumor viruses such as SV40, polyoma, adeno and herpes. In these cases differential gene expression is mainly controlled by transcription initiation from stage-specific promoters.

The best cases of post-transcriptional control of viral gene expression might be found in lentiviruses with the human immunodeficiency virus 1 (HIV-1) (Fig. 6) being the most extensively studied example (Kim *et al.*, 1989). Both the CaMV and HIV-1 genomes contain RNA-based *cis* elements inhibitory to expression, and in both cases transactivators are required to alleviate the inhibition. In CaMV RNA the inhibitory leader sequences and the

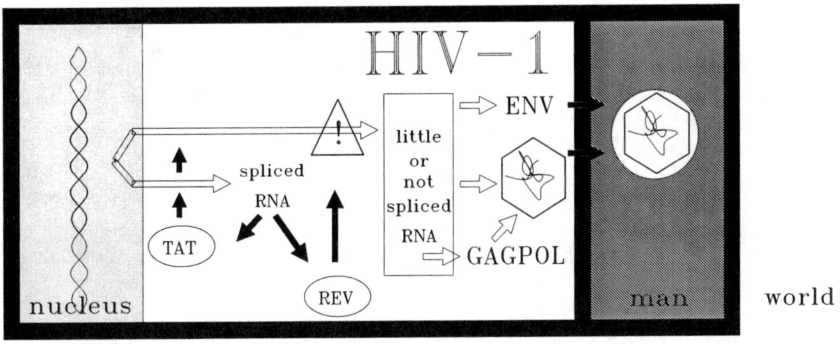

Fig. 6 Regulation of HIV-1 expression. Without transactivation little RNA is produced. In the presence of transactivator (Tat) transcription is increased 20-fold and translation from the resulting RNA 5-fold. RNA produced originally is spliced, giving rise to mRNA for the transactivators Tat and Rev. The Rev transactivator now inhibits splicing or enhances transport of unspliced RNA and once spliced Env-RNA to the nucleus. As a consequence full size RNA can act both as mRNA for the Gag and Gagpol proteins and Env-RNA for envelope protein; proteins and RNA can assemble to form virus particles.

polycistronic arrangement of coding regions require transactivation on the translational level; in HIV-1 splicing signals must be circumvented by rapid transport of full-length RNA or by splicing inhibition with the help of the Rev transactivator (Felber *et al.*, 1988). Additional transactivation is provided by the action of the Tat transactivator on the RNA based Tar sequence, increasing production of transcripts 20-fold and their translation 5-fold (review: Sharp & Marciniak, 1989). In both virus systems protein and virion production depends ultimately on transactivation, but only CaMV RNA appears to be constitutively produced and transported. The reason for this difference might be a direct consequence of the lack of an integration system in CaMV and the accumulation of minichromosomes instead. While a single integrated copy of HIV-1 DNA would guarantee distribution of the virus genome to the daughter cells, many more copies of CaMV minichromosomes must accumulate to prevent loss due to cell proliferation, mutation and perhaps defense mechanisms acting on DNA. The general inhibition of protein production in both cases, however, might be related to the evasion of defense reactions triggered by the proteins, immune response in the HIV-1 case and hypersensitivity in the CaMV case.

ACKNOWLEDGEMENTS

We highly acknowledge the stimulating discussions with Mónica Torruella and the expert technical assistance of Hanny Schmid-Grob, Gundula Pehling and Mathias Müller. S.K. had an IAESTE and M.dT. an EMBO fellowship.

REFERENCES

Baughman, G. and Howell, S. H. (1988). Cauliflower mosaic virus 35S leader region inhibits translation of downstream genes. *Virology* **167**, 125-135.

Bonneville, J-M., Fütterer, J., Sanfaon, H. and Hohn, T. (1989). Posttranscriptional transactivation in Cauliflower Mosaic Virus. *Cell,* in press.

Bonneville, J. M., Hohn, T. and Pfeiffer, P. (1988). Reverse transcription in the plant virus, cauliflower mosaic virus, pp. 23-42 in RNA Genetics. [E. Domingo, J. J. Holland, P.Ahlquist (eds)],CRC Press.

Covey, S. N., Lomonossoff, G. P. and Hull, R. (1981). Characterization of cauliflower mosaic virus DNA sequences which encode major polyadenylated transcripts. *Nucleic Acids Res.* **24**, 6735-6747.

Covey, S. N. *et.al.*, (1990). In press.

Felber, B. K., Hadzopoulou-Cladaras, M., Cladaras, C., Copeland, T. and Pavlakis, G. N. (1988). rev protein of HIV-1 affects the stability and transport of viral mRNA. *Proc.Nat.Acad.Sci.USA* **86**, 1495-1499.

Fütterer, J., Gordon, K., Bonneville, J. M., Sanfaon, H., Pisan, B., Penswick, J. and Hohn, T. (1988). The leading sequence of caulimovirus large RNA can be folded into a large stem- loop structure. *Nucleic Acids Res.* **16**, 8377-8390.

Fütterer, J., Gordon, K., Pfeiffer, P., Sanfaon, H., Pisan, B., Bonneville, J. M. and Hohn, T. (1989). Differential inhibition of downstream gene expression by the cauliflower mosaic virus 35S RNA leader. *Virus Genes.* In press.

Gordon, K., Pfeiffer, P., Fütterer, J. and Hohn, T. (1988). *In vitro* expression of cauliflower mosaic virus genes. *EMBO J.* **7**, 309-317.

Gowda, S., Wu, F. C., Scholthof, H. and Shepherd, R. J. (1989). Gene VI of figwort mosaic virus (caulimovirus group) functions in posttranscriptional expression of genes of the full-length RNA transcript. *Proc.Nat.Acad.Sci.USA.*

Guilley, H., Dudley, R. K., Jonard, G., Balazs, E. and Richards, K. (1982). Transciption of cauliflower mosaic virus DNA: detection of promoter sequences, and characterization of transcripts. *Cell* **30**, 763-773.
Hull, R. (1984). A model for expression of CaMV nucleic acid. *Plant Mol.Biol.* **3**, 121-125.
Jacks, T. and Varmus, H. E. (1985). Expression of the Rous sarcoma pol gene by ribosomal frameshifting. *Science* **230**, 1237-1242.
Kim, S., Byrn, R., Groopman, J. and Baltimore, D. (1989). Temporal aspects of DNa and RNA synthesis during human immunodeficiency virus infection: evidence for differential gene expression. *J.Virol.* **63**, 3708-3713.
Kirchherr, D., Albrecht, H., Mesnard, J-M. and Lebeurier, G. (1988). Expression of the cauliflower mosaic virus capsid gene *in vivo*. *Plant Mol.Biol.* **11**, 271-276.
Lawton, M. A., Tierny, M. A., Nakamura, I., Anderson, E., Komeda, Y., Dub, P., Hoffman, N., Fraley, R. T. and Beachy, R. N. (1987). Expression of a soybean beta-conglycinin gene under the control of the cauliflower mosaic virus 35S and 19S promoters in transformed petunia tissues. *Plant Mol.Biol.* **9**, 315-324.
Maule, A. J. (1985). Replication of Caulimoviruses in plants and protoplasts, pp. 161-190 in Molecular Plant Virology (J. Davies, ed.) CRC Press.
Maule, A. J., Harker, C. L. and Wilson, I. G. (1989). The pattern of accumulation of cauliflower mosaic virus-specific products in infected turnips. *Virology* **169**, 436-446.
Menissier, J., Lebeurier, G. and Hirth, L. (1982). Free cauliflower mosaic virus supercoiled DNA in infected plants. *Virology* **117**, 322-328.
Penswick, J., Hübler, R. and Hohn, T. (1988). A viable mutation in cauliflower mosaic virus separates its capsid protein- and polymerase genes. *J.Virol.* **62**, 1460-1463.
Pfeiffer, P. and Hohn, T. (1983). Involvement of reverse transcription in the replication of cauliflower mosaic virus: a detailed model and test of some aspects. *Cell* **33**, 781-789.
Pietrzak, M. and Hohn, T. (1987). Translation products of Cauliflower mosaic virus ORF V, the coding region corresponding to the retrovirus pol gene. *Virus Genes* **1**, 83-96.
Plant, A. L., Covey, S. N. and Grierson, D. (1985). Detection of a subgenomic mRNA for gene V, the putative reverse transcriptase gene of cauliflower mosaic virus. *Nucleic Acids Research* **13**, 8305-8321.
Rollo, F. and Covey, S. N. (1985). CaMV DNA persists as supercoiled forms in cultured turnip cells. *J.gen.Virol.* **66**, 603-608.
Schultze, M., Hohn, T. and Jiricny, J. (1989). Mode of expression from the cauliflower mosaic virus reverse transcriptase gene. Braunschweig Symp.Appl.Plant Mol.Biol. 323-328.
Sharp, P. A. and Marciniak, A. (1989). HIV TAR: An RNA enhancer?. *Cell* **59**, 229-230.

DISCUSSION OF T. HOHN'S PRESENTATION

R. Ranu: If the GUS gene is moved downstream to the positions of genes II, III or IV, what is the level of synthesis?

T. Hohn: We are not this far with this analysis. However, in constructs with a dimer CAT GUS gene both are expressed.

R. Ranu: What is the mechanism of gene VI in transactivation?

T. Hohn: This transactivator has to be present in large amounts and thus is not like ordinary regulators. The inclusion body organelle makes its own rules on translation by concentrating ribosomes and by holding the RNA in the correct form.

R. Ranu: Does the gene VI product bind to the ribosome rather than to the transcript?

T. Hohn: It binds to the polysome.

Gene VI of Figwort Mosaic Virus Activates Expression of Internal Cistrons of the Full-length Polycistronic RNA Transcript

Siddarame Gowda, Fang C. Wu, H. Scholthof and R. J. Shepherd

Department of Plant Pathology, University of Kentucky,
Lexington, KY 40546, USA

Figwort mosaic virus (FMV) is a caulimovirus with a circular double-stranded DNA genome of about 8 kilobase pairs (kbp) (Richins *et al.*, 1987). Like CaMV, the genome is expressed as two sets of genes (Gowda *et al.*, 1989). The first set consists of a group of six genes (VII and I through V) closely spaced on a major polycistronic transcript which spans the entire viral genome. The second set consists of gene VI which has its own promoter and is transcribed as a separate monocistronic transcript (Fig. 1). This suggests that gene VI may have some early function in the replication cycle. The product of this gene of FMV accumulates in viral inclusion bodies, similar to those of cauliflower mosaic virus (CaMV), the type member of caulimovirus group. Although the role of gene VI of CaMV in disease induction and the host range determination is well documented (Daubert *et al.*, 1984; Schoelz *et al.*, 1986)), its molecular function has only recently been studied. As presented here and by Hohn *et al.*, for CaMV in this volume, as well as a previous report (Hohn *et al.*, 1989), gene VI appears to activate the expression of internal genes of the polycistronic transcript. This role has been demonstrated by the electroporation of viral genome-reporter gene constructs into plant protoplasts. For example, a plasmid containing the promoter of the major RNA transcript, the long intergenic region and gene VII (Fig. 2) followed by the chloramphenicol acetyltransferase (CAT) gene as a downstream cistron, shows little CAT activity following electroporation into plant protoplasts unless a separate plasmid expressing gene VI of FMV is included in the electroporation mixture. When the latter is included during electroporation the internal CAT cistron is expressed at a high level. Various types of mutagenesis of gene VI of the activating plasmid suggest that it is the product of gene VI that is active in the enhanced expression of internal cistrons.

RESULTS AND DISCUSSION

The full length RNA transcript of FMV starts at position 6940 (Fig. 2) and spans the complete genome. This transcript has a 5' leader consisting of 560 nucleotides before the start of the major open reading region (gene VII) at nucleotide 7500 (Fig. 2). Five or six small translational reading regions, speci-

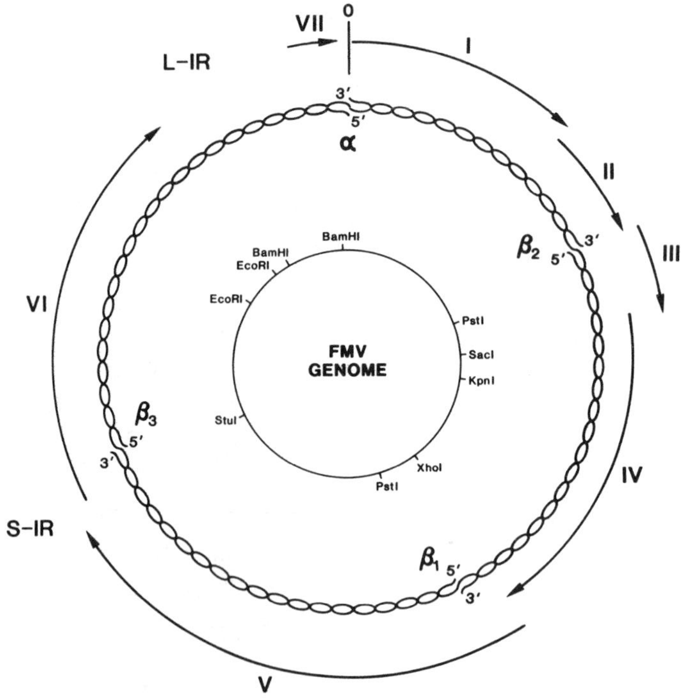

Fig. 1 Physical map of the FMV genome (Gowda *et al.*, 1989). Circular double stranded DNA of the virus is indicated by the interwoven line. It contains 4 interruptions, one in the minus strand (α) and three in the coding strand (β_1, β_2 and β_3). Location of the open reading frames are indicated by the peripheral arrows. Small (S-IR) and the large (L-IR) intergenic regions are also pointed out.

fying from 3 to 21 amino acids, intervene between the 5' end and the AUG start codon of gene VII. These probably account for the repression of translation of gene VII as well as downstream cistrons on the full-length transcript (Baughman and Howell, 1988; Hohn *et al.*, 1989; Futterer *et al.*, 1988). As proposed by Kozak (1986, for a review) a fraction of scanning 40S ribosomal subunits probably drop off the mRNA at the stop codon of each successive open reading region.

The approach used to study the expression of the internal cistrons of the full length RNA transcript consisted of engineering a reporter gene (chloramphenicol acetyltransferase, CAT) into various positions downstream from gene VII (Fig. 2). These DNA constructs were then electroporated into protoplasts of suspension cell cultures of *Nicotiana edwardsonii,* a systemic host for FMV, and transient assays for CAT activity were carried out.

```
CAGCTGGCTTGTGGGGACCAGACAAAAAAGGAATGGTGCAGAATTGTTAG  6739
GCGCACCTACCAAAAGCATCTTTGCCTTTATTGCAAAGATAAAGCAGATT  6789
CCTCTAGTACAAGTGGGGAACAAAATAACGTGGAAAAGAGCTGTCCTGAC  6839
AGCCCACTCACTAATGCGTATGACGAACGCAGTGACGACCACAAAAGAAT  6889
TCCCTCTATATAAGAAGGCATTCATTCCCATTTGAAGGATCATCAGATAC  6939
TGAACCAATATTTCTCACTCTAAGAAATTAAGAGCTTTGTATTCTTCAAT  6989
GAGAGGCTAAGACCCTAAAGAGTTTCGAAAGAGAAATGTAGTATAGTAAG  7039
AGTCCTCCCAGTCCGGGAGATTGTAATAAAGAGATCTTGTAATGGATCCA  7089
AGTGTCTGTAATTTTTGGAAAAATTGATCTATAAAATATTCAATCTTTCT  7139
TTAAGCTTATTCAAAGAACAAACATACTATCTATCATCCAAATCCACAGA  7189
GTGACAGAGAGAAAATGGTCTGTGTTGTGTGGATCTGAAGTACCGCCGAG  7239
GCAGGAGGCCGTTAGGGAAAAAGGGACTGTTTTGACCGTCAAAGTATCAG  7289
GCTGGCTCTAGGAAGGAAGATGAAGATATCAGGTATTGGTTTATGTTCTA  7339
AAAAATAAGTAATAAAGAAAAAAGTTTATTAAAAAGAAAATTTTATCAAG  7389
AGCAAATTACATGTCTAGAGGATACCTAGATCTATATTACAATAATCTTA  7439
CTTACATGTTTTATTTCGTGACTCTAAATTAAAAAATTGTTTAATTGTTT  7489
ATTCAAAACAATGCCAGGACTAACCCTCCAGCAAGAGTATATACTCTTAG  7539
CACACCTTATTCTTCAGGTACTCGAAGAAGTCAAGCAGGTACAACTGCAT  7589
TCAGGAGACTTCCAGTTTCTCAGAAGTCTATATGCTAGGCTTAACGGGCT  7639
TCGGTCACACCAAGCTCATCTCCAAGCGAGAATTTCAGCTGTTTCTCAAC  7689
ACGGCAATTGAAGCGTCATGAACTTCAAGAAAATCTCGGATCC         7732
```

Fig. 2 Nucleotide sequence of the 3' end of gene VI (at nucleotide 6902), the large intergenic region and gene VII of FMV. The 'TATA' box, poly(A) signal and a conserved 35 bp sequence are underlined. The boxed codons indicate the start and stop codons of gene VII. The numbers on the right correspond to nucleotide sequence numbers of the FMV genomic map (Gowda *et al.*, 1989).

When the CAT gene was placed immediately downstream of the promoter for the major RNA transcript (Gowda *et al.*, 1989) (pFMV20 CAT) CAT expression was much higher than when the CAT gene was situated as a separate cistron downstream of the promoter plus the large intergenic region followed by gene VII (pFMV32 CAT; see Fig. 6). This shows that sequences of the large intergenic region followed by gene VII greatly reduced CAT activity when the latter occurred as an internal cistron downstream from the long 5' leader of the full length RNA transcript . Presence of even short stretches of the intergenic region 3' to the polyadenylation signal reduces CAT expression. Others have observed a similar decrease in the activity of the reporter genes downstream of the intergenic region (Baughman and Howell 1988; Hohn *et al.*, 1989). When pFMV32 CAT was electroporated along with a full length infectious clone of FMV (pFMV Sc3) (cloned at its unique Sac I site in region IV leaving gene VI intact) considerable higher CAT activity resulted, compared to electroporation of pFMV32 CAT alone. However, when pFMV32 CAT was electroporated with a full length clone of the severe strain of FMV, (pFMV m3), where the cloning vector interrupts gene VI, no increase in CAT activity was observed (data not presented). This suggested the involvement of gene VI in transactivation of pFMV32 CAT. However, the possibility of the involvement of genes VII and I-III which are present intact on the activating plasmid was not ruled out. Therefore, two gene VI clones were constructed in which gene VI was placed under its own promoter/termination sequences, pFMV RVI (Fig. 3A), and in which gene VI was placed between the 35S promoter of CaMV and the rubisco 3/ terminator in the plasmid pGS1 RVI (Fig. 3B). A 6- and 9-fold increase in CAT activity was observed when pFMV32 CAT was coelectroporated with these plasmids suggesting that gene VI alone was sufficient to transactivate pFMV32 CAT (Fig. 4, lanes 8 and 9).

In order to obtain evidence as to whether gene VI protein was required for transactivation of pFMV32 CAT, two gene VI mutants truncated from the 3' end were generated, i.e. pGS1 RVI Δ *Bgl* II, and pGS1 RVI Δ *Nsi* I. Portions of the 3' end of the gene were removed by using unique restriction enzyme sites, i.e. *Bgl* II located approximately in the middle of the gene, and a unique *Nsi* I site towards the 3' end of the gene. Upon coelectroporation these mutants did not support transactivation, suggesting that the intact protein or RNA transcript is required for enhanced expression (Fig. 5, lanes 12 and 13).

Gene VI of FMV is capable of coding for a 54-KDa protein (Richins *et al.*, 1987) and it contains four in-frame ATG codons in its coding region, all of them in a favorable context for translational initiation. To obtain further evidence for the requirement for the protein of gene VI for transactivation, mutants were prepared with altered ATG's. These were then tested for their ability to transactivate pFMV32 CAT. When the first ATG was changed, the resulting plasmid, pGS1 RVI M1, still retained nearly 75% of its ability to transactivate (Fig. 5 lane 14). However, when the first and the second ATG's were changed, the plasmid PGS1 RVI M2 had lost 75% of its capacity to transactivate. Finally when the first three ATG's were changed the resulting plasmid, pGS1 RVI M3, lost most of its ability to transactivate pFMV32 CAT (Fig. 5, lanes 14-16). These observations suggest that the protein of gene VI, rather than the RNA, participates in transactivation.

The large intergenic region of FMV (Fig. 2) is nearly 600 bp long and contains some of the regulatory elements of the major RNA transcript (Gowda *et al.*, 1989). In order to understand whether 5' leader sequences of the RNA are required for transactivation, plasmids with varying lengths of the leader region were constructed and compared with pFMV32 CAT for transactivation by gene VI. The 5' end of these constructs starts at nucleotide 6690 of the FMV genome which is in the 3' end of gene VI. The latter has been observed to contain the enhancer elements of the promoter for the full length RNA transcript (Gowda *et al.*, 1989). Gene VI-containing plasmids failed to transactivate these truncated leader mutants as they do pFMV32 CAT (Fig. 6). This indicates that the sequence for gene VII is required for the transactivational response.

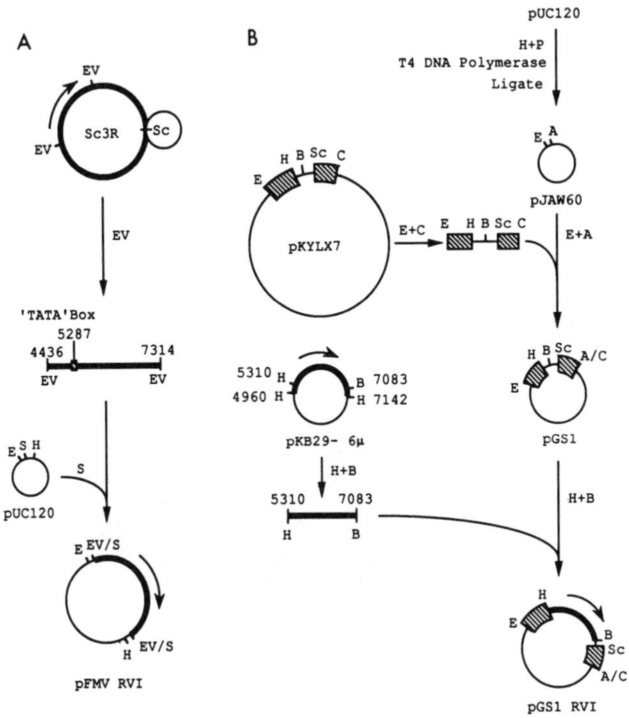

Fig. 3 Construction of the FMV region VI transactivating plasmids pFMV RVI and pGS1 RVI. (A) Plasmid pFMV RVI has region VI of FMV is under the control of its homologous promoter and termination sequences. This plasmid was constructed by cloning the *Eco* RV fragment [(positions 4436-7314 of the FMV genomic map (Richins *et al.* 1987)] into the *Sma* I site of pUC119. (B) pGS1 RVI contains the coding region of FMV gene VI between the CaMV 35S promoter and a rubisco gene 3' terminator sequence (hatched boxes). The latter was constructed as follows: A segment of DNA containing the CaMV 35S promoter and a rubisco gene 3' termination sequence from the transformation vector pKYLX 7 (Schardl *et al.*, 1987) was cut with *Eco* RI and *Cla* I and cloned into pJAW60 (a derivative of pUC119 with the *Pst* I-*Hind* III restriction sites deleted in the polylinker region at the *Eco* RI-*Acc* I window. The resulting plasmid pGS1 was digested with *Hind* III and *Bam* HI and ligated with the *Hind* III and *Bam* HI fragment of pKB29-6m. The latter was generated by creating a *Hind* III restriction site at position 5310 (of the FMV genomic map) by oligonucleotide mutagenesis of pKB29. pKB29 was constructed by cloning a *Hind* III fragment of the FMV genome (4960-7142) into pUC119. The thin solid line represents the vector sequences and the thick solid line indicates viral sequences. The arrow indicates the direction of translation of gene VI and the numbers correspond to nucleotide sequence numbers on the FMV genomic map. Abbreviations used for restriction enzymes: A, *Acc* I; B, *Bam* HI; C, *Cla* I; E, *Eco* RI; EV, *Eco* RV; H, *Hind* III; P, *Pst* I; S, *Sma* I; and Sc, *Sac* I; A/C, *Acc* I/*Cla* I fusion and EV/S; *Eco* RV-*Sma* I fusion.

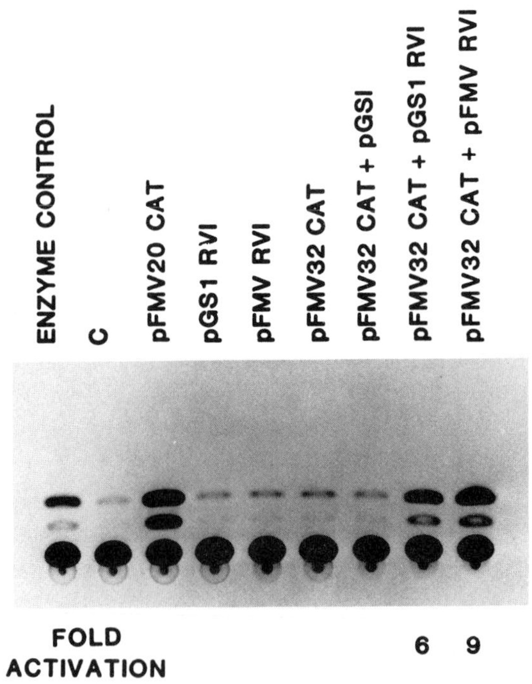

Fig. 4 Transactivation of pFMV32 CAT by FMV gene VI containing plasmids pGS1 RVI and pFMV RVI. 20 μg of the CAT plasmid and 50 μg of the transactivating plasmids were mixed and coelectroporated into protoplasts of suspension cells of *N. edwardsonii*. 24 hours after electroporation, 2 X 10^5 protoplasts were used to assay CAT activity. The plasmid pFMV20 CAT with the CAT gene positioned immediately downstream from the promoter of the full length genomic transcript (without 5' leader and gene VII) was used as a positive CAT expression plasmid to monitor effectiveness of electroporation. (From Gowda *et al.*, 1989).

In the preceding experiments, enhancement of CAT expression was observed when a CAT-containing plasmid and a gene VI construct were mixed for coelectroporation. However, to determine whether or not gene VI *in cis* is adequate for transactivation, a partially redundant clone of a naturally occurring deletion mutant of FMV (Scholthof *et al.*, 1986) containing the CAT gene placed at the 3' end of the fused gene IV/V of this FMV deletion mutant (pH44-3) was constructed (Fig. 7). This plasmid contains an uninterrupted gene VI *in cis*. Upon electroporation the plasmid pH44-3 expressed a high level of CAT activity. However, when the promoter for the major RNA transcript was deleted, the resulting plasmid expressed significantly lower CAT activity (Fig. 7; pH44-3 versus pH44 Δ E). This result indicates that the full length polycistronic RNA transcript is used for expression of the CAT

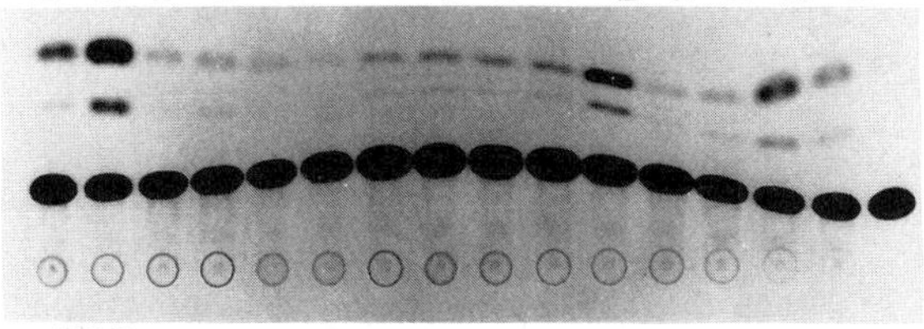

Fig.5 Effect of mutagenesis of gene VI on its effectiveness during transactivation. The gene VI mutants represented in the figure are: pGS1 RVI Δ*Bgl* II-the transactivating plasmid was truncated with *Bgl* II which cleaves in approximately the middle of gene VI; pGS1 RVI Δ*Nsi* I-the transactivating plasmid was truncated with *Nsi* I which removes the 3' terminal 300 nucleotides of gene VI; pGS RVI M1-the first methionine codon (ATG) of gene VI at nucleotide 5363 was mutated to CGG; pGS RVI M2-the first and second methionine codons of gene VI (the latter at nucleotide 5430) were mutated (i.e., the second ATG was changed to AGA); pGS1 RVI M3-the first three methionine codons (the third at nucleotide 5748) were mutated (i.e., the third ATG was changed to GTC). (From Gowda *et al.*, 1989).

gene. Similarly, deletion of the promoter for gene VI reduced CAT activity considerably (Fig. 7, pH44-3 versus pH44 Δ S), showing that transcription of gene VI is necessary for the expression of CAT activity of pH44-3. (Fig. 7). This suggestion was confirmed by our recent observation that coelectroporation of pH44 Δ S and gene VI of FMV (pGS1 RVI) showed significantly higher CAT activity. Other experiments with plasmids containing the CAT gene in gene II or in the start of gene IV, show that in the absence of gene VI *in cis*, the level of CAT expression is significantly decreased (data not presented). The efficient expression of the CAT gene in these far downstream positions suggests that gene VI interacts with viral RNA or the translational apparatus to give either internal initiation or a more efficient re-initiation during modified scanning behavior by eukaryotic ribosomes.

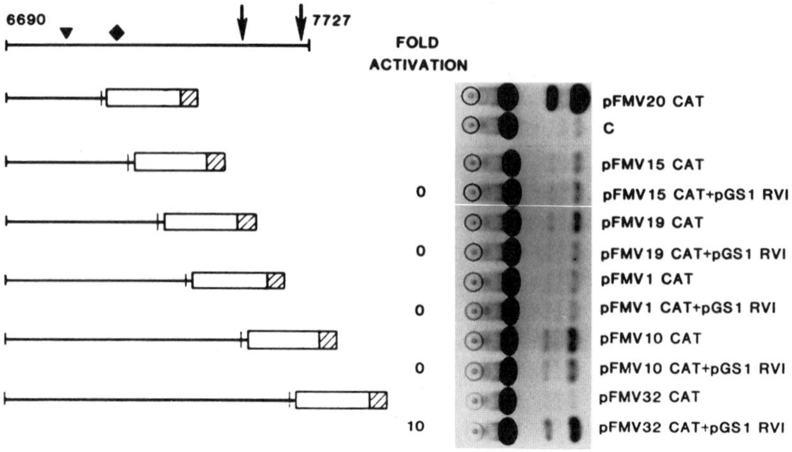

Fig. 6 Comparison of the transactivation of pFMV32 CAT with plasmids containing truncated 5' leader, by gene VI of FMV (pGS1 RVI). The solid line at the top represents the genetic map of a portion of the FMV genome and indicates the positions of 'TATA' box (∇) poly (A) signal (◇) and the start and stop codons (↓↓) of gene VII (also see Fig. 2). The six clones with various portions of the intergenic region are shown below the genetic map. The viral sequences are represented by the narrow solid lines. The open boxes represent the CAT reporter gene with a rubisco gene 3' terminator (hatched area) ligated to its 3'-terminus. All clones start at nucleotide 6690 of the FMV genome (200 bp upstream from the 3'-end of gene VI, see Fig. 1). Clone pFMV20 CAT extends from that point to nucleotide 7003; pFMV15 CAT extends to nucleotide 7105. pFMV19 CAT extends to nucleotide 7223; pFMV1 CAT extends to nucleotide 7316; pFMV10 CAT extends to nucleotide 7504; pFMV32 CAT extends to nucleotide 7667. FMV gene VII starts at nucleotide 7499 and ends at nucleotide 7700 (Richins *et al.*, 1987). The results of the CAT assays are shown in the right portion of the figure. (From Gowda *et al.*, 1989).

These experiments with a second caulimovirus confirm the report by Hohn *et al.*, (1989) for transactivation of reporter genes placed in internal positions downstream from the large intergenic region and gene VII of the CaMV genome. Hence, post-transcriptional transactivation appears to be a general phenomenon for the caulimoviruses. This suggests that gene VI has an early role during the infection process by these viruses. Moreover, our experiments indicate that the sequence of gene VII is required in addition to the protein of gene VI for the transactivational response. CAT gene constructs with other portions of the long intergenic region of FMV do not show enhancement of CAT expression. Whether the gene VII sequence is a site on the RNA transcript that interacts with the gene VI protein or whether it is active *in trans* has not been determined.

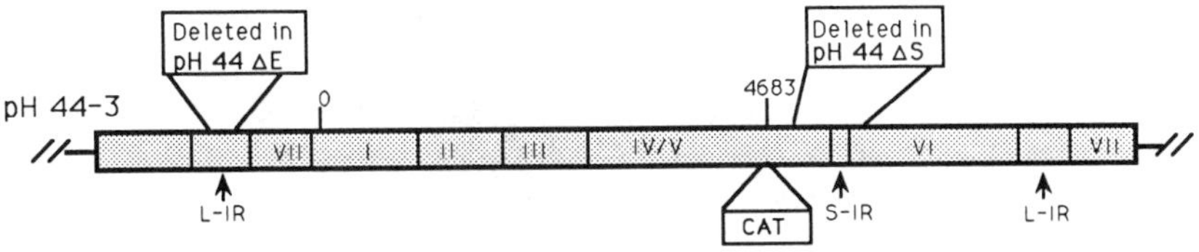

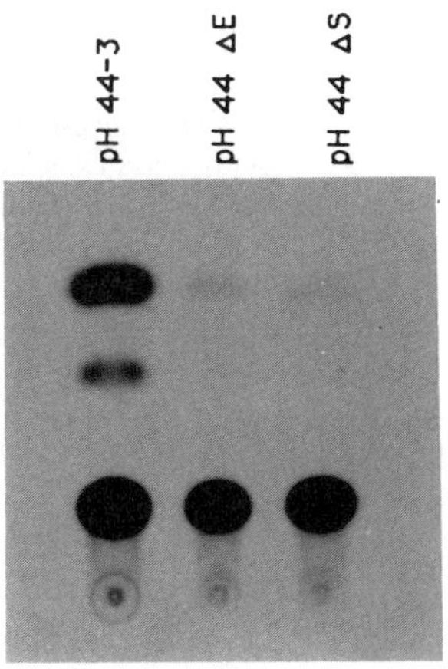

Fig. 7 (Upper) Schematic representation of pH44-3, pH 44ΔE and pH44ΔS. The plasmids are derived from pFHS which is a clone of a naturally occurring deletion mutant of FMV in which most of ORF IV and the 5' end of ORF V is deleted and fused in frame to ORF V to give ORF IV/V. The partially redundant clone shown in the diagram is cloned into pUC 119 and the CAT gene is inserted at position 4683 on FMV genomic map. As a result, a stop codon is introduced in IV/V followed immediately by the start codon for CAT. In pH44ΔE, a 405 bp Eco R1 fragment is deleted which contains sequences essential for the promoter for the full-length RNA transcript. In pH44ΔS, a 584 bp fragment containing the promoter for the gene VI transcript and the 5' end of gene VI is deleted. Single lines represent the vector sequences, the shaded box indicates the viral genome with the ORF'S CAT gene is shown as a separate box. (Lower) The CAT expression of pH44-3 (lane 1), pH44ΔE (lane 2) and pH44ΔS (lane 3), (50 μg each) in *N. edwardsonii* suspension cell protoplasts.

REFERENCES

Baughman, G., Howell, S.H. (1988). Cauliflower mosaic virus 35S RNA leader region inhibits translation of downstream genes. *Virology* **167**:125-135.

Daubert, S.D., Schoelz, J., Li, D. and Shepherd, R.J. (1984). Expression of disease symptoms in cauliflower mosaic virus genomic hybrids. *Jour. Mol. Appl. Genet.* **2**:537-547.

Futterer, J., Gordon, K., Bonneville, J.M., Sanfacon, H., Pisan, B., Penswick, J., Hohn, T. (1988). The leading sequence of caulimovirus large RNA can be folded into a large stem-loop structure. *Nucleic Acids Res* **16**:8377-8390.

Gowda, S., Wu, F.C., Shepherd, R.J. (1989). Identification of promoter sequences for the major RNA transcripts of figwort mosaic virus and peanut chlorotic streak virus. *J Cell Biochem* **13**-D (supp.) 301.

Gowda, S., Wu, F. C., Scholtof, H. B., Shepherd, R. J. (1989). Gene VI of figwort mosaic virus (caulimovirus group) functions in transcriptional expression of genes on the full-length RNA transcript. *Proc. Natl. Acad. Sci. USA* **86**: 9203-9207

Hohn, T., Bonneville, J.M., Futterer, J., Gordon, K., Pisan, B., Sanfacon, H., Schultze, M., Jiricny, J. (1989). The first thousand and one nucleotides of genomic CaMV RNA. In Molecular Biology of Plant-Pathogen Interactions. UCLA Symposium on Molecular and Cellular Biology, (eds. Saskawicz, B., Ahlquist, P., Yoder, O.) Alan R. Liss. New York. pp.153-165.

Kozak, M. (1986). Point mutations define a sequence flanking the AUG initiator codon that modulates translation by eukaryotic ribosomes. *Cell* **44** : 283-292.

Richins, R.D., Scholthof, H.B., Shepherd, R.J. (1987). Sequence of figwort mosaic virus DNA (caulimovirus group) *Nucleic Acids Res* **15**:8451-8466.

Schardl, C.L., Byrd, A.D., Benzion, G., Altschuler, M.A., Hildebrand, D.F. and Hunt, A.G. (1987). Design and construction of a versatile system for the expression of foreign genes in plants. *Gene* **61**:1-11.

Schoelz, J.E., Shepherd, R.J., Daubert, S. (1986). Region VI of cauliflower mosaic virus encodes a host range determinant. *Molec Cell Biol* **6** : 2632-2637.

Scholthof H., Richins R.D., Handley M. K., Shepherd R. J., (1986). Nucleotide sequence of a naturally occuring deletion mutant of figwort mosaic virus. *Phytopathology* **76**:1131.(Abst.)

DISCUSSION OF S. GOWDA'S PRESENTATION

R. Hull: Have either of you done binding studies between gene VI products and any of the caulimovirus nucleic acids?

T. Hohn: No, but in the case of CaMV the gene VI product has a putative zinc finger and a basic region at its C-terminus. Both zinc and RNA are bound but we have not yet studied the specificity of the binding.

S. Gowda: We have not yet studied this aspect.

Molecular Genetics of Tomato Golden Mosaic Virus Replication: Progress Toward Defining Gene Functions, Transcription Units and the Origin of DNA Replication

David M. Bisaro, Garry Sunter, Gwen N. Revington*, Clare L. Brough, Sheriar G. Hormuzdi, and Marcos Hartitz

The Biotechnology Center and Department of Molecular Genetics.
The Ohio State University, Columbus, OH 43210, USA

The geminiviruses are a unique group of infectious agents that replicate and cause disease in a wide variety of plant species. Because of their small DNA genomes and their ability to multiply to high copy number, these viruses have the potential to serve as important model systems for the processes of transcription and DNA replication in their hosts. Geminiviruses also have attracted considerable attention due to their potential to be modified for use as amplifiable vectors for stable plant transformation and transient expression assays. As a result, a great deal of effort has been directed toward understanding the molecular biology of these pathogens, and in the past few years considerable progress has been made with several group members. Many excellent reviews documenting this progress are available (Stanley, 1985; Harrison, 1985; Davies *et al.*, 1987; Lazarowitz, 1987; Davies and Stanley, 1989). As this paper is based primarily on recent work done in this laboratory and on earlier work done in collaboration with the laboratory of Dr. Stephen G. Rogers, it will focus mainly on tomato golden mosaic virus and other bipartite geminiviruses. However, pertinent information about other group members will be discussed where appropriate.

GEMINIVIRUS CHARACTERISTICS

The geminiviruses take their name from the unusual structure of their capsids, which are composed of two fused, incomplete T=1 icosahedra whose interiors are thought to be connected. The paired particles measure ~18 X 30 nm, are constructed from a single 27-34 kD coat protein species and contain a single molecule of covalently closed circular single-stranded DNA. The single-stranded circles vary in size between 2.5-3.0 kb, depending on the virus.

The geminivirus group can be divided into two subgroups on the basis of

* Present address: Department of Botany and Microbiology, Auburn University, Auburn, AL 36849

insect vector and genome type. One subgroup is comprised of viruses which are transmitted by leafhoppers and have genomes composed of a single molecule of DNA. Members of this subgroup infect mono- or dicotyledonous plants and include maize streak virus (MSV), wheat dwarf virus (WDV), beet curly top virus (BCTV), digitaria streak virus (DSV), and chloris striate mosaic virus (CSMV). All of these viruses replicate only in phloem parenchyma cells and are either difficult to transmit by conventional mechanical methods or are transmissible only through leafhoppers. However, it is possible to infect plants with these viruses using the "agroinfection" technique (Grimsley *et al.*, 1986; 1987; Grimsley and Bisaro, 1987), which exploits the T-DNA transfer system of *Agrobacterium tumefaciens* to deliver infectious viral DNA into susceptible cells. Briefly, this technique involves the construction of Ti plasmids containing tandemly repeated dimers or partial dimers of viral DNA in the T-DNA. In plant cells, unit-length viral genomes capable of initiating an infection are released from the T-DNA either by recombination or by a single-strand replication mechanism.

Members of the second geminivirus sub-group are restricted to dicot hosts and are transmitted in nature by the whitefly. These viruses have genomes divided between two DNA molecules of similar size which are both required for infectivity (Hamilton *et al.*, 1983; Stanley, 1983). The bipartite geminiviruses include tomato golden mosaic virus (TGMV), cassava latent virus (CLV, also known as ACMV: African cassava mosaic virus), bean golden mosaic virus (BGMV), and squash leaf curl virus (SqLCV). Like their monopartite counterparts, whitefly-borne geminiviruses are considered to be phloem-limited although some, including CLV and TGMV, also infect mesophyll cells (Sequeira and Harrison, 1982; Rushing *et al.*, 1987). The bipartite viruses usually can be transmitted mechanically as virus particles, as isolated single- or double-stranded DNA, or as cloned DNA following excision from the cloning vector. Cloned DNAs also are transmissible as uncut plasmids if the components are cloned as tandem dimers (Hayes *et al.*, 1988a), although inoculation with TGMV is most efficient when agroinfection is employed (Elmer *et al.*, 1988a, Hayes *et al.*, 1988b).

GEMINIVIRUS GENOME ORGANIZATION

The genomes of several bipartite geminiviruses have been cloned and sequenced. The two genomic DNAs of TGMV (Bisaro *et al.*, 1982; Hamilton *et al.*, 1984), CLV (Stanley and Gay, 1983), and BGMV (Howarth *et al.*, 1985) differ considerably in nucleotide sequence except for a highly conserved common region of approximately 230 nucleotides. Although common regions are not conserved between viruses, a 33-35 nucleotide inverted repeat sequence found in all geminiviruses is located in this region and within the analogous region of monopartite group members. The inverted repeat is capable of forming a hairpin structure (Sunter *et al.*, 1985) with a 12-13 nucleotide loop sequence that is nearly identical in all group members.

A map of the TGMV genome is presented in Fig. 1. It contains six open reading frames (ORFs) which are conserved with respect to size and relative location in the genomes of other bipartite geminiviruses, and these are positioned on both strands of double-stranded DNA intermediates that are synthesized in infected cells. The 2588 bp TGMV DNA A contains four ORFs which diverge from the common region. The lone rightward ORF (AR1) encodes the 28.7 kD viral coat protein (Kallendar *et al.*, 1988). Leftward ORFs include AL1 (40.3 kD), AL2 (14.9 kD) and AL3 (15.7 kD). Genome component B (2522 bp) has two coding regions, BR1 (29.3 kD) and BL1 (26.4 kD), which also diverge from the common region. It should be pointed out, however, that so far only the viral coat protein and AL1 protein (J. S. Elmer and S. G. Rogers, personal communication) have been identified in extracts from infected plants. The existence and properties of the remaining viral proteins are inferred from their ORFs and by the behavior of mutants containing lesions within them.

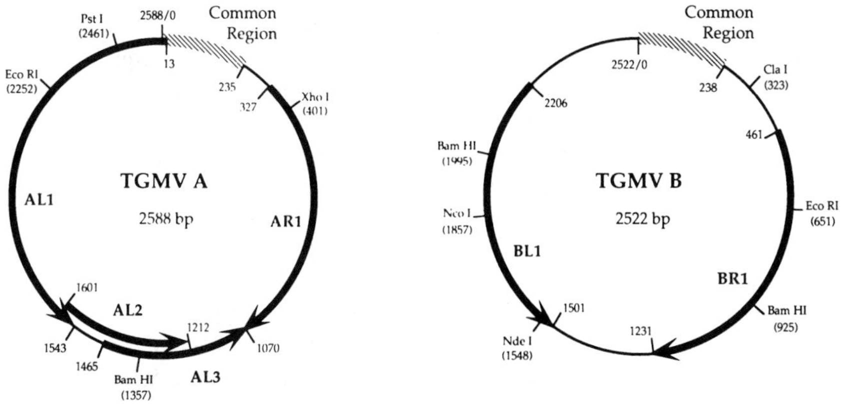

Fig. 1 Maps of TGMV A and B genome components. The double-stranded RF DNAs of TGMV are drawn from the sequence of Hamilton *et al.*, (1984) as modified by MacDowell *et al.*, (1986). The solid arrows define the positions of open reading frames and the hatched boxes indicate the common region. Positions of restriction sites are indicated outside the circles . (From Sunter and Bisaro, 1989, by permission).

Experiments with plants transgenic for tandemly repeated copies of TGMV A and B DNAs have allowed us to determine the molecular basis for the mutual dependence of the two genome components. These studies have shown that DNA A encodes all viral functions necessary for the replication and encapsidation of viral DNA, while DNA B provides functions required for movement of the virus or viral DNA through the infected plant (Rogers *et al.*, 1986; Sunter *et al.*, 1987). Similar conclusions were reached by Stanley and co-workers, who obtained complementary results using protoplasts inoculated with CLV genome components (Townsend *et al.*, 1986).

TGMV GENE FUNCTION

Mutagenesis of TGMV ORFs has suggested the general role of each coding region in the multiplication cycle. TGMV mutants have been constructed in each ORF using a variety of techniques, including restriction endonuclease-mediated deletion, repair and ligation of ends created by restriction endonuclease digestion, linker insertion, and site-directed mutagenesis. Mutant viral genomes were cloned into Ti plasmid vectors and introduced into *Nicotiana benthamiana* plants or leaf discs by agroinfection. In experiments with B component mutants, plants or leaf discs were co-inoculated with the mutant and the wild-type A component. Infectivity of A component mutants was assessed by inoculating plants transgenic for the B component.

The results of genetic studies are summarized in Table 1. The viral coat protein, which is encoded by ORF AR1, is not necessary for infectivity and may be deleted, although coat protein mutants display attenuated and delayed symptom development (Gardiner *et al.*, 1988). This finding has prompted the replacement of the coat protein sequence with bacterial coding sequences (e.g. chloramphenicol acetyltransferase, neomycin phosphotransferase and β-glucuronidase), with the result that infected plants systemically express these activities from the viral coat protein promoter. To date, foreign gene expression has been achieved with vectors of similar design in inoculated plants (Hayes *et al.*, 1988c; Ward *et al.*, 1988; Rogers *et al.*, 1989), in transgenic plants containing containing chromosomal, tandemly repeated "master copies" of the modified viral genome (Hayes *et al.*, 1988c; Hayes *et al.*, 1989), and in a transient leaf disc system (Hanley-Bowdoin *et al.*, 1988). In the case of inoculated plants, there appears to be some preference for the replication and/or spread of viral genomes that are wild-type in size, and this may cause problems with vector stability. However, as TGMV molecules containing relatively large insertions are stable in leaf discs and transgenic plants, it appears that size selection is imposed only when cell-cell virus movement is required. This implies that there is an optimal size, ~ 2.5 kb, at which viral DNA most efficiently replicates and passes between cells.

Further studies of the A component coding regions have demonstrated that ORF AL3 is not essential for systemic infection, although mutants show attenuated symptoms that are similar in appearance to those produced by coat protein deletion mutants. Using a leaf disc DNA replication assay, it has been shown that AL3 mutants replicate all forms of viral DNA (Elmer *et al.*, 1988b), however, analysis of AL3 mutants in transfected protoplasts indicates that the amount of DNA they produce is reduced relative to wild-type controls (manuscript in preparation). In contrast to AR1 and AL3, both the AL1 and AL2 ORFs are required for infectivity. In leaf discs and transfected protoplasts, AL2 mutants produce double-stranded viral DNA in amounts similar to wild-type controls but accumulate only reduced amounts of single-stranded DNA. AL1 mutants, however, do not replicate DNA in leaf discs or in protoplasts, and hence the product of this ORF is necessary for the replication of all forms of viral DNA (Elmer *et al.*, 1988b).

Table 1. Genetic Analysis of TGMV ORFs

ORF	Size(kD)	Mutant symptoms	TGMV DNAs* in: plants	discs/protoplasts	Function
AR1	28.7	attenuated	ds, ss	ds, ss (reduced)	coat protein
AL1	40.3	none	none	none	required for replication
AL2	14.9	none	none	ds, ss (reduced)	movement, ss DNA levels
AL3	15.7	attenuated	ds, ss	ds (reduced), ss (reduced)	DNA replication levels
BR1	29.3	none	none	ds, ss	movement
BL1	26.4	none	none	ds, ss	movement

Observations, conclusions and conjectures drawn from mutational analysis of TGMV ORFs are summarized. Replication of A component mutants in *N. benthamiana* plants, leaf discs and protoplasts was assessed by Southern blot hybridization. The replication of B component mutants in protoplasts has not been studied.
*ds, double-stranded DNA; ss, single-stranded DNA.

Mutagenesis experiments with DNA B ORFs have shown that the products of BR1 and BL1 are required for infectivity (Brough *et al.*, 1988; Etessami *et al.*, 1988), although both BR1 and BL1 mutants are capable of replicating in agroinfected leaf discs (unpublished observations). It therefore appears that the products of the two B component ORFs as well as the product of ORF AL2 are required either directly or indirectly to effect virus movement through the infected plant. How these proteins might interact with viral DNA, virus particles, host-coded proteins, and each other during the movement process is unknown.

REPLICATION OF TGMV DNA

TGMV, like most geminiviruses, accumulates in the nuclei of infected cells, from which unit-length and concatameric single- and double-stranded viral

DNA forms can be recovered (Hamilton *et al.*, 1982; Rushing *et al.*, 1987; Slomka *et al.*, 1988). The presence of such double-stranded DNA intermediates suggests that TGMV and other geminiviruses employ a multiplication strategy that is similar to the one used by single-stranded DNA-containing coliphages such as ΦX174 and M13. In these phages, host functions convert the infectious single-stranded DNA to a double-stranded replicative (RF) form that is first amplified and later serves as template for both viral transcription and single-strand DNA synthesis. Some additional evidence that this general scheme is used by TGMV comes from direct analysis of replicating DNA in transfected protoplasts (Fig. 2; manuscript in preparation). In our assay system, protoplasts derived from *Nicotiana tabacum* suspension culture cells are transfected with a pUC118 derivative containing one and one-half copies of the A genome component. Following Southern blot analysis of transfected protoplast DNA, supercoiled TGMV DNA can be detected as early as 12 hours post-transfection while single-stranded forms begin to accumulate by 18-24 hours post-transfection.

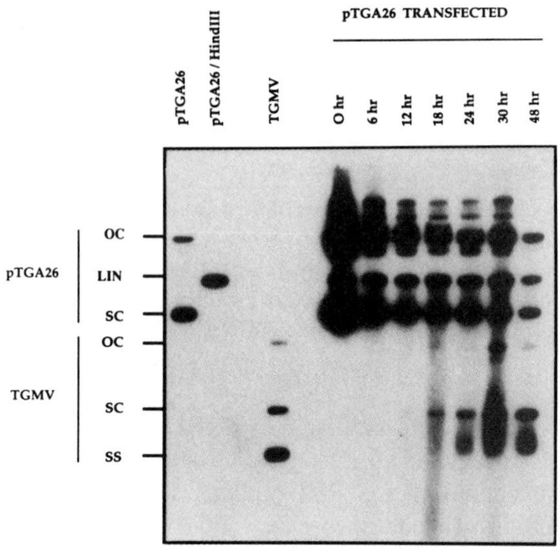

Fig. 2 TGMV DNA replication in *N. tabacum* protoplasts. Protoplasts prepared from suspension culture cells were transfected in the presence of polyethylene glycol (Potrykus *et al.*, 1985) with plasmid pTGA26, which is composed of a TGMV A 1 1/2-mer in pUC118 (manuscript in preparation). pTGA26 contains a single HindIII cleavage site. DNA extracts were prepared from protoplasts harvested at various times post-transfection and subjected to Southern blot analysis using a ^{32}P-TGMV A-specific probe. An autoradiograph of a blot containing ~2 μg of protoplast DNA per lane is shown. The positions of open circular (OC), linear (LIN), supercoiled (SC) and single-stranded (SS) pTGA26 and TGMV DNAs are indicated.

A great deal of circumstantial evidence suggests that the common region contains an origin of replication. Recently, we have obtained direct evidence which indicates that this region is in fact required for the replication of TGMV DNA B and that sequences within and near the conserved inverted repeat are essential for this process (Revington *et al.*, 1989). The B component was chosen for this study because mutations in DNA A might affect replication either by disrupting *cis*-acting sequences or by interfering with the expression of AL1. Mutations in the B common region should permit identification only of *cis*-acting sequences. As shown in Figure 3, the 413 bp PstI fragment containing the

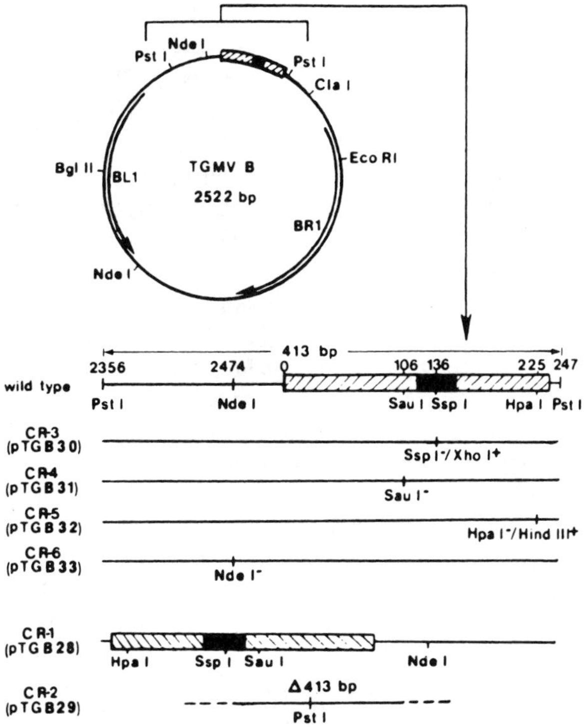

Fig. 3 Diagrams of TGMV B common region mutants. The shaded box within the common region (cross-hatched box) indicates the position of the putative hairpin structure that is found in all geminiviruses. The arrows show the position of ORFs BR1 and BL1. The expanded region represents the 413 bp PstI fragment and restriction sites where alterations within this fragment were constructed for each CR mutant are noted. CR-1 and CR-2 contain an inversion and a deletion of the PstI fragment, and CR-3 and CR-5 were created by insertion of an XhoI linker and a HindIII linker, respectively. CR-4 and CR-6 were constructed by end-fill and ligation of cohesive ends generated by restriction . (From Revington *et al.*, 1989, by permission).

common region of TGMV B was altered by inversion (CR-1), by deletion (CR-2), or by the insertion of a 2-10 bp sequence at the SspI site (CR-3), the SauI site (CR-4), the HpaI site (CR-5) and the NdeI site (CR-6). The resulting B component mutants were co-inoculated with the wild-type A component into healthy *N. benthamiana* plants and leaf discs by agroinfection. Symptoms characteristic of TGMV infection were observed within 21 days in plants inoculated with wild-type DNA B as well as in plants inoculated with mutants CR-5 and CR-6 (data not shown). Plants inoculated with CR-4 also became infected but displayed delayed and attenuated symptoms. Subsequent analysis of DNA extracts showed that plants and leaf discs infected with CR-5 and CR-6 contained DNA B in amounts comparable to plants inoculated with the wild-type B component (Fig. 4). Plants and leaf discs inoculated with the attenuated mutant CR-4 contained reduced amounts of B DNA, while those inoculated with non-infectious mutants (CR-1, CR-2 and CR-3) did not contain this genome component in detectable amounts. However, the presence of TGMV B in plants co-inoculated with TGMV A and mutants CR-1 and CR-3 may be inferred by the small amounts of A DNA in plants inoculated with these mutants (Fig. 4). This suggests that movement factors provided by TGMV B are required only in very small quantities, and raises the possibility that the movement factors themselves may be transported between cells.

The results are consistent with a TGMV B replication origin residing within the common region. Removal of the common region by a 413 nucleotide deletion (CR-2) abolishes the molecule's ability to be replicated. Likewise, inversion of the 413 nucleotide sequence (CR-1), or insertion of 8 nucleotides into the loop of the conserved hairpin structure (CR-3) reduces replication to a level below the limit of detection in our assay system. While the result with CR-1 is somewhat puzzling, it is possible for the inversion mutation to affect viral DNA replication if initiation is strand-specific or if *cis*-acting sequences which enhance or are required for the process are critically spaced within and outside of the inverted region. Further, the end points of the inversion might lie within a sequence which is required for the normal binding of replication factors. The insertion of an XhoI linker into the SspI site within the putative hairpin loop (mutation CR-3) alters this sequence so that it has the potential to form a more stable structure (G = -42.6 vs. -31.6 kcal/mol). In the hypothetical mutant structure, 8 of the 12 nucleotides which form the loop in the wild-type configuration are base paired. As the hairpin structure and its loop sequence are highly conserved in all geminiviruses, our observation corroborates the hypothesis that it plays an important role in viral DNA replication. The insertion within the SauI site (CR-4) reduced the amount of B DNA replicated. This sequence, which lies about 30 nucleotides upstream from the hairpin loop, must therefore also be important for the efficient replication of TGMV B. In contrast, insertions about 90 nucleotides downstream of the hairpin loop (CR-5) or 180 nucleotides upstream of it (CR-6), had no detectable effect on the ability of TGMV B to replicate.

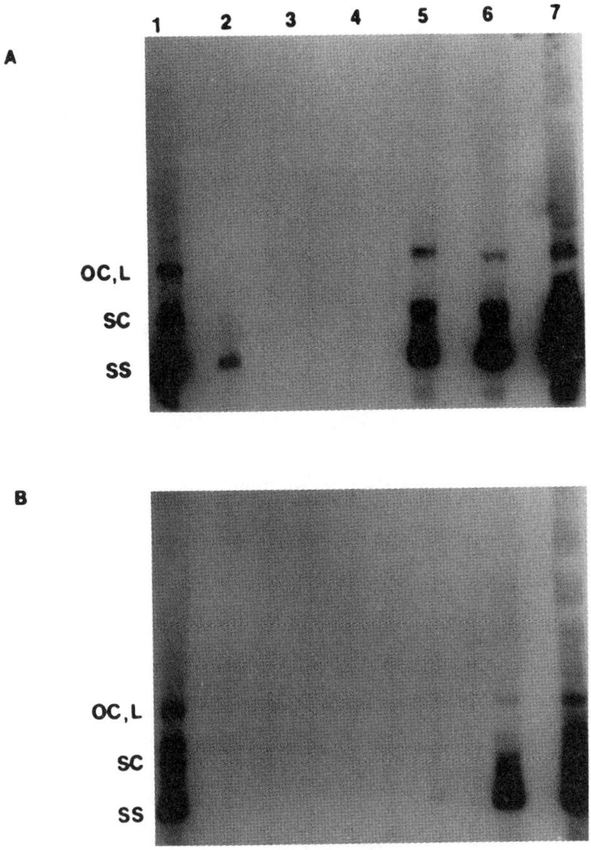

Fig. 4 Southern blot analysis of DNA isolated from tobacco plants inoculated with wild-type TGMV A and TGMV B common region mutants. (A) TGMV A-specific probe. (B) TGMV B-specific probe. Lanes contain DNA from plants inoculated with TGMV A and the following B components: Lane 1, wild type TGMV B (3 μg); lane 2, CR-1 (10 μg); lane 3, CR-2 (10 μg) ; lane 4, CR-3 (10 μg) ; lane 5, CR-4 (10 μg); lane 6, CR-5 (3 μg); lane 7, CR-6 (5 μg). Lane 8 contains marker TGMV DNA. The positions of single-stranded (SS) and double-stranded open circular (OC), linear (L) and supercoiled (SC) forms of TGMV DNA are indicated . (From Revington *et al.*, 1989, by permission).

As the common region is nearly identical in TGMV A and B, it probably fulfills similar functions in both components. Therefore, it is likely that sequences identified as necessary or important for the replication of B DNA also should be important for the replication of A DNA. Experiments are in progress to define further the replication origin of TGMV B and to identify a minimal sequence that will confer replication competence to a heterologous molecule in the presence of the A component.

TGMV TRANSCRIPTION

In the bipartite viruses, ORFs diverge from the common regions of both genome components. Consistent with this genome organization, polyadenylated virus-specific RNAs have been mapped to both strands of both components of CLV (Townsend *et al.*, 1985) and TGMV (Sunter *et al.*, 1989). The coat protein transcripts of both viruses have been mapped more precisely by S_1 and primer extension analysis. In TGMV, the 860 nucleotide coat protein transcript begins at nucleotide 319 and ends in the vicinity of nucleotide 1090 (Petty *et al.*, 1988; Sunter *et al.*, 1989). As the AR1 translation initiation codon lies at nucleotide 327, the RNA has an untranslated leader sequence of only 7 or 8 nucleotides. This is in marked contrast to the relatively large (~160 nucleotide) untranslated leader of the CLV coat protein transcript, which also contains four apparently non-functional AUGs before the initiation codon. The significance of the difference in leader size and of the additional AUGs is unknown at present.

The 920 nucleotide TGMV BR1 transcript also has been mapped (Sunter and Bisaro, 1989). This RNA has its 5' end located between nucleotides 445-451, a major 3' end near nucleotide 1250 and two minor 3' ends near nucleotides 1340 and 1375. Comparison of the TGMV AR1 and BR1 transcription units reveals some interesting similarities (Fig. 5). Both map to similar positions on their respective genome components and neither appears to be processed. Further, the AR1 and major BR1 polyadenylation signals both partially overlap their translation stop codons. The two RNAs have short untranslated leader sequences of 7-8 (AR1) and 13-15 (BR1) nucleotides, and their 5' termini are ~30 nucleotides downstream of a consensus TATAA sequence and ~65 (BR1) and ~95 (AR1) nucleotides downstream of a consensus CAAT box sequence.

Northern blot analysis of polyadenylated RNA from infected plants has indicated that the leftward, complementary sense transcripts of TGMV DNA A belong to three size classes (1.6, 1.0, and 0.7 kb) which appear to be 3' co-terminal (Sunter *et al.*, 1989). Leftward transcription from the B component results in a single RNA size class of ~1.3 kb. However, upon closer inspection, we have found the leftward transcription units of both TGMV DNAs to be more complex than indicated by initial low resolution mapping (Sunter and Bisaro, 1989). The results of S_1 analysis and primer extension experiments indicate that the 1.6 kb AL1 and the 1.3 kb BL1 RNA species are composed of a family of transcripts with distinct 5' ends. The AL1 RNAs have 5' termini which map to nucleotides 62, 2548, 2540 and 2515, while the 5' ends of BL1 RNAs map to nucleotides 62, 2449 and 2237/2240 (Fig. 5). Three of the A-specific RNA start sites have been reported previously (Hanley-Bowdoin *et al.*, 1988). We are currently mapping the 1.0 and 0.7 kb AL2/AL3 transcripts in order to determine if these also are more complex than our original experiments have indicated.

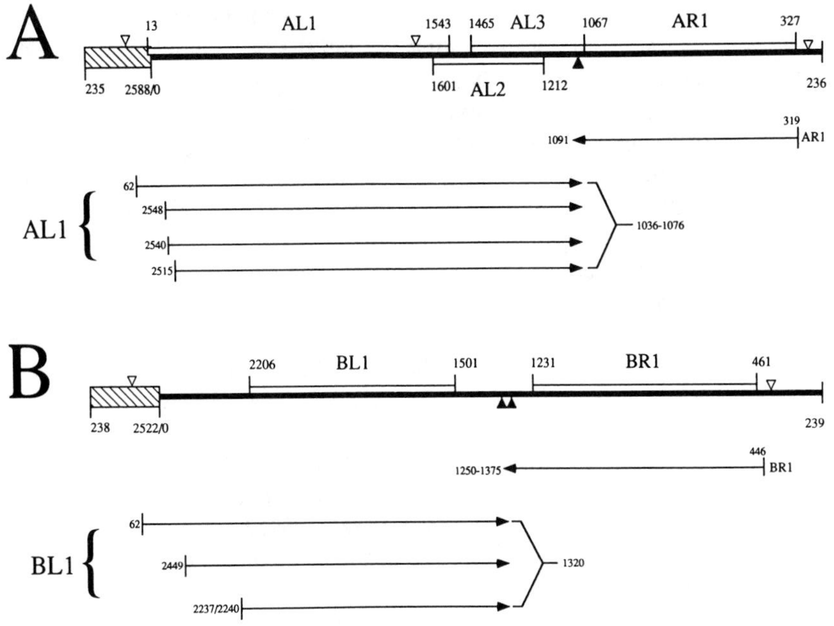

Fig. 5 Map of TGMV B and partial map of TGMV A transcription units. Data obtained from primer extension and S_1 nuclease analysis of TGMV transcripts are summarized (Sunter *et al.*, 1989; Sunter and Bisaro, 1989). The diagram shows, in linear form, the map positions of A and B component transcripts (arrows) relative to the common region (hatched boxes), A and B component ORFs, TATAA sequences (open triangles) and polyadenylation signals (shaded triangles). The complementary sense 1.0 kb and 0.7 kb RNAs detected by Northern blot analysis are not shown. (From Sunter and Bisaro, 1989, by permission).

The BL1 RNAs are 3' co-terminal near nucleotide 1320 and appear to share a common polyadenylation signal at nucleotide 1339. The AL1 RNAs may share a consensus polyadenylation signal located at nucleotide 1088, ~12 nucleotides upstream of an apparent 3' end at nucleotide 1076. A second 3' end maps to nucleotide 1036 and is located ~20 nucleotides downstream of a non-consensus (AATAAT) polyadenylation signal. Considered together, the data indicate that there is 50-75 nucleotides of overlap at the 3' ends of the viral and complementary sense transcription units of TGMV DNAs A and B. The significance of transcription overlap is unknown, but it may have some role in the regulation of viral gene expression. S_1 nuclease analysis of 5' and 3' ends has not provided evidence for the involvement of RNA processing in the generation of these transcripts. Both the AL62 and BL62 RNAs lie 30 nucleotides downstream of a consensus TATAA sequence, while the remainder of the RNAs are not preceded by a recognizable promoter element.

The significance of the multiple complementary sense RNAs is unclear. On the A component, AL62 is the only RNA that can encode the entire 40.3 kD AL1 ORF. The remaining RNAs initiate within the N-terminal portion of this coding sequence, which is highly conserved among the geminiviruses and is necessary for viral DNA replication. Monopartite viruses that infect monocots possess overlapping ORFs of ~30 kD and 17 kD, which correspond to the amino and carboxy terminus of AL1, respectively (Mullineaux *et al.*, 1985). It has been reported that splicing of a WDV (Schalk *et al.*, 1989) and a DSV (Accotto *et al.*, 1989) transcript occurs in such a way that the two ORFs are fused to produce one which resembles AL1. This processing event is required for WDV replication in protoplasts but it is not known whether the unspliced RNAs themselves are translated. It is possible that one or more of AL2548, AL2540 and AL2515 RNAs are used for translation of the C-terminal portion of AL1, perhaps beginning at a methionine codon at nucleotide 2013. Alternatively, the RNAs may be used for the translation of ORFs AL2 and AL3, although the 1.0 kb and 0.7 kb complementary sense RNAs map closer to the initiation codons of these ORFs. Another possibility is that the transcripts may be used for the translation of ORFs which are presently unrecognized. This also may be the case for BL1 RNAs, which all initiate upstream of the translation initiation codon of the 26.4 kD BL1 ORF.

DISCUSSION

Mutational analysis of the TGMV genome has allowed the general role of each ORF in the viral replication cycle to be determined. These studies have revealed that AL1 is the only viral gene product which is absolutely required for viral DNA replication. The product of the AL2 ORF appears to be necessary for the accumulation of single-stranded TGMV DNA, since AL2 mutants produce reduced amounts of this form while viral double-stranded DNA forms accumulate in infected cells to amounts comparable to wild-type virus. However, maximal replication of both single- and double-stranded DNA forms depends upon the presence of a functional AL3 gene product. The movement-defective phenotype exhibited by AL2 mutants implies that it is single-stranded DNA that is transported between cells in mutants which can not produce virus particles. As BR1 and BL1 mutants are not infectious and do not appear to affect viral DNA synthesis, the products of these ORFs may play more direct roles in cell-cell spread of the virus and/or viral DNA .

Work on the elucidation of viral gene function has now reached the point where it is necessary to initiate studies of the biochemical nature of viral proteins. With the exception of coat protein, this will require the synthesis of TGMV polypeptides in heterologous expression systems. Purified viral proteins can be used to raise antibody and will be necessary to carry out *in vitro* biochemical assays using defined components and/or plant cell extracts. In this way, it will be possible to determine the location of viral proteins in infected cells, and ultimately to learn the molecular nature of the interactions which take place between virus and host proteins during the infection process.

The information we have obtained from transcript and origin mapping studies suggests that the ~230 bp intergenic regions of TGMV DNAs contain promoters and replication origin sequences which control or are necessary for transcription and DNA replication. Specifically, the hairpin-loop structure conserved in all geminiviruses is necessary for viral DNA replication, and promoters undoubtably are located near transcription start sites that map ~70 bp upstream and ~180 bp (DNA A) or ~325 bp (DNA B) downstream of the inverted repeat. Thus, the structure of the TGMV genome resembles that of SV40, and invites comparison of these viruses. The ~5.0 kb SV40 chromosome also is organized such that transcripts diverge from a relatively small intergenic region (~375 bp) which contains promoter and enhancer sequences necessary for transcription as well as inverted repeats that are part of the "core" origin of replication. SV40 enhancer sequences can compensate for one another to some extent, and also exert some control over the activity of the replication origin in addition to their role in the regulation of viral transcription. The enhancer sequences in turn depend on interactions with host and viral proteins for their activity (DePamphilis, 1988). Although no direct evidence for functional homology between viral proteins and regulatory sequences exists at this time, it will be interesting to compare and contrast the regulatory schemes of SV40 and TGMV as more information concerning TGMV replication and transcription becomes available. Any discussion of replication and control of gene expression, however, also must take into account the single-stranded phase of the geminivirus multiplication cycle, which appears similar in many respects to the single-stranded phages.

In conclusion, it is becoming clear that geminiviruses have much to offer both as model systems for the study of plant gene expression and as vectors to facilitate the study of heterologous plant genes, as has been the case for SV40 in mammalian cells. Like the single-stranded phages, geminiviruses also hold considerable promise as important model systems for the study of plant DNA replication.

ACKNOWLEDGEMENTS

This work was supported in part by USDA grant 87-CRCR-1-2541, NSF grant DMB-8796326 and by a grant from the Monsanto Company.

REFERENCES

Accotto, G. P., Donson, J., and Mullineaux, P. M. (1989). Mapping of digitaria streak virus transcripts reveals different RNA species from the same transcription unit. *EMBO J.* **8**, 1033-1039.

Bisaro, D. M., Hamilton, W. D. O., Coutts, R. H. A., and Buck, K. W. (1982). Molecular cloning and characterization of the two DNA components of tomato golden mosaic virus. *Nucleic Acids Res.* **10**, 4913-4922.

Brough, C. L., Hayes, R. J., Morgan, A. J., Coutts, R. H. A., and Buck K. W. (1988). Effects of mutagenesis *in vitro* on the ability of cloned tomato golden mosaic virus DNA to infect *Nicotiana benthamiana* plants. *J. Gen. Virol.* **69**, 503-514.

Davies, J. W., Townsend, R., and Stanley, J. (1987). The Structure, Functions and Possible Exploitation of Geminivirus Genomes. *In* "Plant Gene Research: Plant DNA Infectious Agents" (T. Hohn and J. Schell, Eds.), pp. 31-52. Springer-Verlag, Vienna.

Davies, J. W., and Stanley, J. (1989). Geminivirus genes and vectors. *Trends in Genetics* **5**, 77-81.

DePamphilis, M. L. (1988). Transcriptional elements as components of eukaryotic origins of DNA replication. *Cell* **52**, 635-638.

Elmer, J. S., Sunter, G., Gardiner, W. E., Brand, L., Browning, C. K., Bisaro, D. M., and Rogers, S. G. (1988a). *Agrobacterium*-mediated inoculation of plants with tomato golden mosaic virus DNAs. *Plant Mol. Biol.* **10**, 225-234.

Elmer, J. S., Brand, L., Sunter G., Gardiner, W. E., Bisaro, D. M., and Rogers, S. G. (1988b). Genetic analysis of tomato golden mosaic virus II. The product of the AL1 coding sequence is required for replication. *Nucleic Acids Res.* **16**, 7043-7060.

Etessami P., Callis, R., Ellwood S., and Stanley, J. (1988). Delimitation of essential genes of cassava latent virus DNA 2. *J. Gen. Virol.* **16**, 4811-4829.

Gardiner, W. E., Sunter, G., Brand, L., Elmer, J. S., Rogers, S. G., and Bisaro, D. M. (1988). Genetic analysis of tomato golden mosaic virus: the coat protein is not required for systemic spread or symptom development. *EMBO J.* **7**, 899-904.

Grimsley, N., Hohn,T., and Hohn, B. (1986). Agroinfection, an alternative route for plant virus infection using Ti plasmid. *Proc. Natl. Acad. Sci. USA* **83**, 3282-3286.

Grimsley, N., Hohn,T., Davies, J. W., and Hohn, B. (1987). *Agrobacterium*-mediated delivery of infectious maize streak virus into maize plants. *Nature* **324**, 177-179.

Grimsley, N., and Bisaro D. (1987). Agroinfection. *In* "Plant Gene Research: Plant DNA Infectious Agents" (T. Hohn and J. Schell, Eds.), pp.88-107. Springer-Verlag, Vienna.

Hamilton, W. D. O., Bisaro, D. M., Coutts, R. H. A., and Buck, K. W. (1982). Identification of novel DNA forms in tomato golden mosaic virus infected tissue. Evidence for a two component viral genome. *Nucleic Acids Res.* **10**, 4901-4912.

Hamilton, W. D. O., Bisaro, D. M., Coutts R. H. A., and Buck, K. W. (1983). Demonstration of the bipartite nature of the genome of a single-stranded DNA plant virus by infection with the cloned DNA components. *Nucleic Acids Res.* **11**, 7387-7396.

Hamilton, W. D. O., Stein V. E., Coutts, R.H.A., and Buck K. W. (1984). Complete nucleotide sequence of the infectious cloned DNA components of tomato golden mosaic virus: potential coding regions and regulatory sequences. *EMBO J.* **3**, 2197-2205.

Hanley-Bowdoin, L., Elmer, J. S., and Rogers, S. G. (1988). Transient expression of heterologous RNAs using tomato golden mosaic virus. *Nucleic Acids Res.* **16**, 10511-10528.

Hayes, R. J., Brough, C. L., Prince, V. E., Coutts, R. H. A., and Buck, K. W. (1988a). Infection of *Nicotiana benthamiana* with uncut cloned tandem dimers of tomato golden mosaic virus DNA. *J. Gen. Virol.* **69**, 209-218.

Hayes, R. J., Coutts, R. H. A., and Buck, K. W. (1988b). Agroinfection of *Nicotiana* spp. with cloned DNA of tomato golden mosaic virus. *J. Gen. Virol.* **69**, 1487-1496.

Hayes, R. J., Petty, I. T. D., Coutts, R. H. A., and Buck, K. W. (1988c) Gene amplification and expression in plants by a replicating geminivirus vector. *Nature* **334**, 179-182.

Hayes, R. J., Coutts, R. H. A., and Buck, K. W. (1989). Stability and expression of bacterial genes in replicating geminivirus vectors in plants. *Nucleic Acids Res,* **17**, 2391-2403.

Harrison, B. D. (1985). Advances in geminivirus research. *Annu. Rev. Phytopathol.* **23**, 55-82.

Howarth, A. J., Caton J., Bossert, M., and Goodman, R.M. (1985). Nucleotide sequence of bean golden mosaic virus and a model for gene regulation in geminiviruses. *Proc. Natl. Acad. Sci. USA* **82**, 3572-3576.

Kallender, H., Petty, I. T. D., Stein, V. E., Panico, M., Blench, I. P., Etienne, A. T., Morris, H. R., Coutts, R. H. A., and Buck, K. W. (1988). Identification of the coat protein gene of tomato golden mosaic virus. *J. Gen Virol.* **69**, 1351-1357.

Lazarowitz, S. G. (1987). The molecular characterization of geminiviruses. *Plant Mol. Biol. Reporter* **4**, 177-192.

MacDowell, S. W., Coutts, R. H. A., and Buck, K. W. (1986). Molecular characterization of subgenomic single-stranded and double-stranded DNA forms from plants infected with tomato golden mosaic virus. *Nucleic Acids Res.* **14**, 7967-7984.

Mullineaux, P. M., Donson, J., Stanley, J., Boulton, M. I., Morris-Krsinich, B. A. M., Markham, P. G., and Davies, J. W. (1985). Computer analysis identifies sequence homologies between potential gene products of maize streak virus and those of cassava latent virus and tomato golden mosaic virus. *Plant Mol. Biol.* **5**, 125-131.

Petty, I. T. D., Coutts R. H. A., and Buck, K. W. (1988). Transcriptional mapping of the coat protein gene of tomato golden mosaic virus. *J. Gen. Virol.* **69**, 1359-1365.

Potrykus, I., Saul M., Petruska, J., Paszkowski, J., and Shillito, R. (1985). Direct gene transfer to cells of a graminaceous monocot. *Mol. Gen. Genet.* **199**, 183-188.

Revington, G., Sunter, G., and Bisaro, D. M. (1989). DNA sequences essential for replication of the B genome component of tomato golden mosaic virus. *The Plant Cell* **1**, 985-992.

Rogers, S. G., Bisaro D. M., Horsch, R. B., Fraley, R. T., Hoffmann, N. L., Brand, L., Elmer, J. S., and Lloyd, A. M. (1986). Tomato golden mosaic virus A component DNA replicates autonomously in transgenic plants. *Cell* **45**, 593-600.

Rogers, S. G., Elmer, J. S., Sunter, G., Gardiner, W. E., Brand, L., Browning, C. K., and Bisaro, D. M. (1989). Molecular genetics of tomato golden mosaic virus. *In* "Molecular Biology of Plant-Pathogen Interactions" (B. Staskawicz, P. Ahlquist, and O. Yoder, Eds.), UCLA Symposia on Molecular and Cellular Biology, New Series, Vol. 101, pp. 199-216. Alan R. Liss, New York.

Rushing, A. E., Sunter, G., Gardiner, W. E., Dute, R. R., and Bisaro, D. M. (1987). Ultrastructural aspects of tomato golden mosaic virus infection in tobacco. *Phytopathol.* **77**, 1231-1236.

Schalk, H. J., Matzeit, V., Schiller, B., Schell, J., and Gronenborn, B. (1989). Wheat dwarf virus, a geminivirus of graminaceous plants, needs splicing for replication. *EMBO J.* **8**, 359-364.

Sequeira, J. C., and Harrison, B. D. (1982). Serological studies on cassava latent virus. *Ann. Appl. Biol.* **101**, 33-42.

Slomka, M., Buck, K. W., and Coutts, R. H. A. (1988). Characterisation of multimeric DNA forms associated with tomato golden mosaic virus infection. *Arch. Virol.* **100**, 99-108.

Stanley, J. (1983). Infectivity of the cloned geminivirus genome requires sequences from both DNAs. *Nature* **305**, 643-645.

Stanley, J., and Gay, M. (1983). Nucleotide sequence of cassava latent virus DNA. *Nature* **301**, 260-262.

Stanley, J. (1985). The molecular biology of geminiviruses. *Adv. Virus Res.* **30**, 139-177.

Sunter, G., Buck, K. W., and Coutts, R. H. A. (1985). S1-sensitive sites in the supercoiled double-stranded form of tomato golden mosaic virus DNA component B: identification of regions of potential alternative secondary structure and regulatory function. *Nucleic Acids Res.* **13**, 4645-4659.

Sunter, G., Gardiner, W. E., Rushing, A. E., Rogers S. G., and Bisaro, D. M. (1987). Independent encapsidation of tomato golden mosaic virus A component DNA in transgenic plants. *Plant Mol. Biol.* **8**, 477-484.

Sunter, G., Gardiner, W.E., and Bisaro, D. M. (1989). Identification of tomato golden mosaic virus-specific RNAs in infected plants. *Virology* **170**, 243-250.

Sunter, G., and Bisaro, D. M. (1989). Transcription map of the B genome component of tomato golden mosaic virus and comparison with A component transcripts. *Virology* **173**, 647-655.

Townsend, R., Stanley, J., Curson, S. J., and Short, M. N. (1985). Major polyadenylated transcripts of cassava latent virus and location of the gene encoding coat protein. *EMBO J.* **4**, 33-37.

Townsend, R., Watts, J., and Stanley, J. (1986). Synthesis of viral DNA forms in *Nicotiana plumbaginafolia* protoplasts inoculated with cassava latent virus (CLV); evidence for the independent replication of one component of the CLV genome. *Nucleic Acids Res.* **14**, 1253-1265.

Ward, A., Etessami, P., and Stanley, J. (1988). Expression of a bacterial gene in plants mediated by infectious geminivirus DNA. *EMBO J.* **7**, 1583-1587.

DISCUSSION OF D. BISARO'S PRESENTATION

R. Hull: If the coat protein gene of African cassava mosaic virus (ACMV) is taken out the genome refills during replication. If the TGMV coat protein gene is removed is there any evidence for infilling after several passages?

D. Bisaro: We find no evidence for infilling and the deletion mutants are very stably propagated. The reason for this difference is not clear as the experiments were carried out in the same host (*N. benthamiana*). But, as reported from Ken Buck's lab, if large pieces of DNA are inserted in the place of the coat protein gene there are often deletions back to wild type size. It is possible that movement is affected by the genome size and that the ACMV size requirement is more stringent than that of TGMV.

J. Hammond: Have you considered replacing the AR1 reading frame with any of the reading frames from DNA B so that you would get a single component, systemically moving virus?

D. Bisaro: Yes, but we have not done the experiment. One of the problems is the risk of creating a different virus. Another way of doing this would be to inoculate AR1 mutants into plants expressing DNA B genes. This could complement and make a movement proficient virus.

R. Gilbertson: The geminiviruses are very important economically world wide. Have you done any work on TGMV in its natural host, tomato?

D. Bisaro: We find it difficult to infect tomato with TGMV. We have done some work in petunia where it does not appear to have long distance systemic movement but moves from cell to cell. We have tried several cultivars of tomato unsuccessfully. The virus was obtained in 1980 from

Brazil and the isolate we are working with has not replicated in tomato for some time.

R. Hayes: How much is the decrease in replication of the AL3 mutants?

D. Bisaro: This has not been quantified yet but a preliminary estimate is by more than 10-fold.

R. Hayes: As you see variable amounts of single-stranded DNA in AL2 mutants would it not be better to inoculate with an AL2 and coat protein mutant to see if the variability of the single-stranded DNA is due to coat protein?

D. Bisaro: The AL2 mutants do make coat protein and therefore one is looking at the replication in the presence of functional AR1 product. We believe that the effects that we are seeing are due to the mutation in the AL2 protein.

R. Hayes: Have you made any plants transgenic in single open reading frames of DNA A and tried crossing them with plants transgenic in the B genome?

D. Bisaro: No. We are doing that at present.

B. Harrison: Does your work with these mutants give any information on the fibrillar rings (which are actually hollow spheres) found in the nuclei of infected plants?

D. Bisaro: No, we have not looked at any of these mutants by electron microscopy.

SUMMARY AND CONCLUDING REMARKS BY THE SECOND SESSION CHAIRMAN, ROGER HULL

In this session we have been dealing with two groups of viruses which encapsidate DNA. However, the caulimoviruses replicate DNA to RNA to DNA and it is perhaps a quirk of fate that they encapsidate the DNA phase. Geminiviruses replicate DNA to DNA and thus can not be easily compared with that of caulimoviruses, at least at the level of replication.

Two major aspects of caulimoviruses have been covered. The first, that of systemic host and symptom determinants raises some interesting questions as to what really are hosts of these viruses. There is no information as to whether the caulimoviruses studied can not replicate in plant species designated as 'non-hosts' or whether they can replicate at the single cell level and the restriction is on spread from initially infected cells. This information is really needed to further the understanding of the effects of the gene VI product. On the available evidence, and assuming that the virus does not replicate in non-hosts, it would appear that the gene VI product only causes symptoms in non-hosts. This and other information that was presented emphasises the importance of the interactions of the viral and host genomes in pathogenesis.

The second aspect, that of the transactivation of expression brought about by the gene VI product, shows a further commonality between caulimoviruses. It provides a mechanism which answers the criticisms pointed at the 'relay race' hypothesis for the expression of the 5' genes of caulimoviruses and especially the question as to how gene expression is controlled. It also suggests a further function for the gene VI product. It is too early to speculate on any possible correlation between the transactivating properties of the gene VI product and its ability to induce symptoms in non-hosts.

The study of geminiviruses is also at the stage of discovering the functions of gene products. Because of their small size and the possible lower rates of recombination the main approach to this is by mutagenesis. It is most likely that, like caulimoviruses, geminiviruses have inbuilt feedback and transactivating mechanisms between gene products and regions of their genomes. There was some evidence pointing to this presented here and no doubt further evidence will be forthcoming.

Organization and Expression of Potyviral Genes

J.G. Shaw[1], A.G. Hunt[2], T.P. Pirone[1] and R.E. Rhoads[3]

Departments of Plant Pathology[1], Agronomy[2] and Biochemistry[3],
University of Kentucky, Lexington, KY 40546, USA

The potyviruses have long been favored subjects for study by plant virologists. They constitute the most numerous of the three dozen or so groups of plant viruses, and collectively are responsible for more damage to the world's crop plants than is caused by the viruses of most of if not all the other groups. This was rather dramatically illustrated by a recently conducted international election of "favorite" filamentous plant viruses among several eminent virologists in which the potyviruses emerged with a clear "victory" (Milne, 1988). The ecological and epidemiological aspects of diseases caused by potyviruses, with their many fascinating but varied and complex considerations, have also prompted the major efforts in research that have been directed over many years to this group of viruses.

In spite of the importance and undoubted charm of the potyviruses, however, it is only in relatively recent years that details of the structure and mechanisms of expression of their genomes have begun to be understood. Urged on with what seems almost to be the spirit of making up for lost time, virologists are now putting concentrated effort into research on the molecular biology of these viruses. It is, in fact, pleasing to note how old-fashioned seem the not so distant complaints about difficulties in preparing and maintaining purified virus as being major obstacles of our research efforts.

This article will consist of a brief review of some recent research concerning the organization of the potyviral genome and the interesting set of proteins which it expresses. While there is ample reason to be pleased and perhaps amazed at how much information relevant to this subject has been amassed in such a relatively short period, it will be somewhat distressing to record how little we yet really know of the functions of most of the gene products and of how they may contribute to disease development.

Information bearing upon the fascinating mechanisms by which potyviral polyproteins are processed by virus-encoded proteases will not be included here as this is the subject of the following chapter.

ORGANIZATION OF POTYVIRAL GENOMES

One of the most interesting characteristics of potyviruses is the array of non-structural viral proteins, some in the form of complexes of rather exotic structure, which are produced in infected cells. A protein of approximately 70 kDa, the cylindrical inclusion protein (CI), aggregates in the cytoplasm to form intricate pinwheel-like or scroll-shaped structures (Edwardson, 1974; Lesemann, 1988). The nuclei of cells infected with some potyviruses contain large crystalline inclusions made up of two proteins (Knuhtsen *et al.*, 1974). With other potyviruses, these inclusions may be tubular or fibrillar in shape, may be located in the cytoplasm, or may not be apparent in infected cells. In the case of tobacco etch virus (TEV), which does produce crystalline nuclear inclusions, these proteins have long been referred to as the 49K and 58K (formerly, 49K) proteins. However, since the size of the nuclear inclusion proteins seems to vary from one potyvirus to another, the names NIa and NIb were introduced by Domier *et al.*, (1986). Potyviruses also produce a protein of approximately 50 kDa, known as the helper component protein (HC), which mediates the transmission of virus particles by aphids (Pirone and Thornbury, 1984). This protein may be a component of the amorphous inclusions found in the cytoplasm of cells infected with some potyviruses (de Mejia *et al.*, 1985). When to this collection of proteins are added the coat protein and VPg, it is perhaps not surprising that potyviral genomes tend to be larger than those of many other RNA plant viruses.

The genomes of potyviruses are contained in filamentous particles 700-900 nm in length and approximately 11 nm in diameter. Few details of particle structure are available (Tollin and Wilson, 1988), a situation that is understandable but in need of redress. Information concerning the mechanisms by which potyvirus particles are assembled (McDonald and Bancroft, 1977) and disassembled is scanty at best.

Potyvirus particles were reported many years ago to contain single-stranded RNA of molecular weight approximately 3 X 10^6 (Hill and Benner, 1976). The RNA was later shown to contain a genome-linked protein (VPg) at the 5'-terminus (Hari, 1981; Siaw *et al.*, 1985) and to be polyadenylated at the 3'-terminus (Hari *et al.*, 1979).

It is only within the past 10-12 years that studies directed toward determination of the organization of the potyviral genome have been reported. *In vitro* translation of the RNAs of TEV and pepper mottle virus (Dougherty and Hiebert, 1980b) and tobacco vein mottling virus (TVMV; Hellmann *et al.*, 1980) resulted in tentative assignments to particular locations in the genomes of genes encoding some of the known viral proteins. Additional cell-free translation studies of these and several other potyviral RNAs followed (summarized in Hiebert and Dougherty, 1988) and resulted in various versions of a potyviral genetic map. However, with the application of additional techniques, including hybrid-arrested translation and nucleotide sequence analysis, the issue of the order and approximate locations of genes in the potyviral genome was soon settled (Hellmann *et al.*, 1986). A current and probably typical potyviral genetic map, that of TVMV, is shown in Fig. 1.

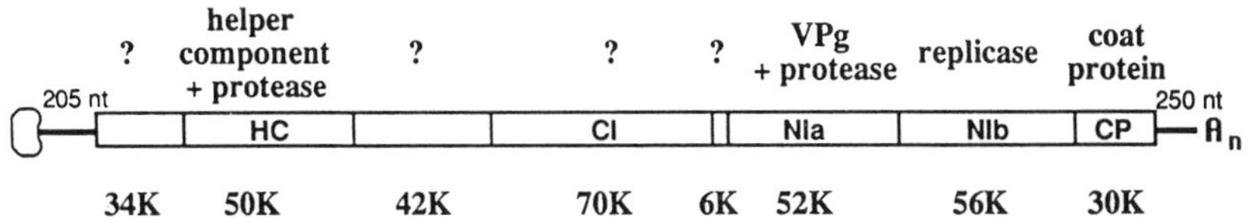

Fig. 1 Gene map of TVMV. Open-box region of diagram represents positions of the putative and known proteins in the polyprotein (amino-terminus at left). HC = helper component protein, CI = cylindrical inclusion protein, NIa and NIb = nuclear inclusion proteins, CP = coat protein. Functions (known and proposed) and approximate sizes (in kDa) of proteins shown above and below boxes, respectively.

STRUCTURE OF POTYVIRAL GENOMES

The complete nucleotide sequences of the RNAs of TVMV (Domier *et al.*, 1986) and TEV (Allison *et al.*, 1986) were reported three years ago. They confirmed several previously proposed aspects of the structure of potyviral RNA and revealed a number of very interesting features concerning mechanisms of genome expression and gene product function. More recently, the nucleotide sequences of two other potyviral RNAs, those of potato virus Y (PVY; Robaglia *et al.*, 1989) and plum pox virus (PPV; Maiss *et al.*, 1989; Lain *et al.*, 1989) have been reported. The sequences of parts of the genomes, in particular the coat protein gene and 3' termini, of a number of other potyviruses are also available.

The genomes of potyviruses are approximately 9.5-9.8 kb in length. Their nucleotide sequences reveal single, long open reading frames of 9.0-9.4 kb capable of encoding proteins of 340-355 kDa. Translation of the RNAs is presumed to be initiated at AUG codons positioned from 145 to 205 nucleotide residues from the 5' termini. Stop codons precede untranslated 3' terminal sequences of 186-250 nucleotides which are terminated by poly(A) tracts of variable length.

Regions of apparently conserved nucleotide sequence are present in the 5'-untranslated region of some potyviral RNAs. Within the first 25 residues from the 5'-termini, there is considerable sequence homology, including a common ACAACAU motif (Turpen, 1989), between the RNAs of TVMV, TEV, PVY and PPV (Fig. 2). An additional conserved sequence, UCAAGCA at nucleotide residue position 43-49 in TVMV and TEV RNAs, is also present in two strains of PVY RNA and this sequence (or one-residue variants therof) is repeated throughout these A-U-rich 5'-proximal regions (Turpen, 1989).

There is considerable variation, from 166 to 475 nucleotide residues (Hammond & Hammond, 1989), in the lengths of the 3'-non-coding regions among the several potyviruses for which the appropriate sequences are available. The potential polyadenylation signal sequence (AAUAAA) has been detected in this region of some but not all potyviral RNAs. However, its functional significance is questionable since, where present, it appears to be further than usual from the site of poly(A) addition.

```
              ** **  * *  **** *******    *

TVMV:     5'-AAAAUAAAACAAAUCAACACAACAUUAUAAC.....

TEV:      5'-NAAAUAACAAAUCUCAACACAACAUAUACAA.....

PVY:       5'-AAUUAAAACAACUCAAUACAACAUAAGAAA.....

PPV       5'-AAAAUAUAAAAACUCAACACAACAUACAAAA.....
```

Fig. 2 Nucleotide sequences at the 5'-termini of the RNAs of TVMV (Domier *et al.*, 1986, 1989), TEV (Allison *et al.*, 1986), PVY (Robaglia *et al.*, 1989) and PPV (Maiss *et al.*, 1989; Lain *et al.*, 1989). * indicates positions in which the same nucleotide is present in all four sequences.

Terminal, non-coding nucleotide sequences, or their complementary sequences, are undoubtedly of importance in initiation of replication and translation and perhaps in the initiation of virion assembly, but direct evidence of participation in such activities is not yet available. Potential stem-loop structures, have been detected in the non-coding termini of some potyviral RNAs (Turpen, 1989), but there do not seem to be highly conserved regions of stable secondary structure.

POTYVIRAL PROTEINS HSUSU

The potyviral genome encodes a long polyprotein which is processed by cis- and transacting viral proteases in co- and post-translational reactions to yield the proteins which are found in infected cells (Carrington and Dougherty, 1987; Hellmann *et al.*, 1988; Dougherty and Carrington, 1988; Carrington *et al.*, 1989a). Six such proteins are known, and analysis of the deduced amino acid sequence of the polyprotein reveals the probable existence of at least two additional potyviral proteins (Table 1).

This section includes brief descriptions of the known and putative potyviral proteins. It must be noted that, with the exceptions of several of the TEV proteins, assignments of the amino- and carboxy-termini of most potyviral proteins have been made on the basis of the locations in the polyproteins of amino acid sequence motifs thought to be protease cleavage sites. There is an obvious and pressing need for verification or correction of these assignments by direct analyses of the termini of these proteins.

Table 1. Some characteristics of potyviral proteins

Protein	Size	Putative Function
"28K/34K Protein"	28-34 kDa	Cell-to-cell movement
Helper Component (HC), HC-Pro	50-56 kDa	Aphid transmission, polyprotein processing (protease)
"42K Protein"	29-42 kDa	Unknown
Cylindrical Inclusion Protein (CI)	ca.70 kDa	Replication (nucleotide binding protein)
"6K Protein"	6 kDa	Unknown
NIa, 49K Proteinase	49-52 kDa	VPg, polyprotein processing (protease)
NIb, 58K Protein	56-58 kDa	Replication (polymerase)
Coat Protein	29-37 kDa	Encapsidation of viral RNA, aphid transmission

"34K Protein"

The existence of an amino-terminal protein of 28 kDa was predicted by the amino acid sequence of the TVMV polyprotein (Domier *et al.*, 1986). (Subsequent information has suggested that this putative protein may be 34 kDa in size but this is by no means certain.) However, such a protein has not yet been found in extracts of infected cells and a mechanism by which it might be released from the polyprotein co- or post-translationally is not known.

A slight degree of amino acid sequence homology has been demonstrated between the putative N-terminal proteins of some potyviruses (Domier *et al.*, 1987; Maiss *et al.*, 1989) and the 30 kDa transport (movement) protein of TMV (Leonard & Zaitlin, 1982; Deom *et al.*, 1987). While it is obviously quite premature to attempt to make a strong case for a cell-to-cell movement function of the N-terminal potyviral protein, there are now available the biochemical and genetic techniques with which to address the question in some detail.

Helper Component Protein (HC)

In natural circumstances, most potyviruses are transmitted from one plant to another by aphids. In order to transmit potyvirus particles, the aphids must have simultaneous or prior access to the HC, a virus-encoded, non-structural protein of approximately 50 kDa (Pirone and Thornbury, 1984; Thornbury *et al.*, 1985). The manner in which this protein mediates transmission is not known but there is some evidence which suggests that it may be involved in binding of virus particles at sites within the aphids. Berger and Pirone (1986) reported that when aphids were allowed to take up virus particles in the presence of HC, the particles accumulated within the maxillary stylets and adjacent tissues; when HC was not provided, virus particles were not detected in the mouthparts of the insects.

Non-transmissibility by aphids of potyviruses such as the potato virus C strain (PVC) of PVY (Govier and Kassanis, 1971) and the PAT strain of zucchini yellow mosaic virus (ZYMV; Lecoq, 1986) appears to be due to lack of production of active HC. In the case of PVC, a polypeptide which reacts with PVY HC antiserum and which co-migrates electrophoretically with PVY HC has been detected by Western blotting (D.W. Thornbury and T.P. Pirone, unpublished data). The amount of this polypeptide produced in PVC-infected plants is comparable with that of PVY HC produced in PVY-infected plants. These observations suggest that relatively minor changes in the HC gene of PVY have resulted in the production of biologically inactive HC by PVC.

As discussed below, lack of, or poor, aphid transmissibility of potyviruses can also be due to defects in the coat protein. The nucleotide sequences of aphid-transmissible and -non-transmissible isolates of PPV are available (Maiss *et al.*, 1989; Lain *et al.*, 1989) but it has not yet been determined whether the lack of transmissibility is due to changes in the HC or the coat protein.

The putative HC regions of the potyviral polyproteins for which amino acid sequence information is available display moderate sequence similarity (45-50% identity). Some potyviruses are transmitted, not by aphids, but by other types of insects or by mites (Murant *et al.*, 1988), and it will be very interesting to compare the amino acid sequences of the HCs of these with the aphid-transmitted potyviruses.

Attention has been drawn to the presence, at or near the putative amino termini of some potyviral HCs, of a zinc-finger motif, the consensus distribution of Cys and His residues found in nucleic acid binding proteins (Sehnke *et al.*, 1989; Robaglia *et al.*, 1989). How such metal-binding domains might be involved in the transmission by aphids of virus particles is not known.

The carboxy-terminal half of the 56-kDa HC of TEV has been shown to contain a proteolytically active domain (Carrington *et al.*, 1989a). This enzyme appears to be a cysteine-type protease and to be responsible for autolytic cleavage of the polyprotein at its carboxy-terminus (Carrington *et al.*, 1989b; Oh and Carrington, 1989). It apparently does not cleave the polyprotein at the amino-terminus of the HC and the activity by which this particular processing event occurs remains unknown. Because of the bi-functional nature of the HC, Carrington *et al.*, (1989a) propose the name HC-Pro for this protein. It will be most interesting to determine whether the proteolytic activity plays a role in the mediation or specificity of aphid transmission of potyviral particles.

"42K Protein"

Examination of the primary structure of the TVMV polyprotein revealed the possible existence of a protein of 42 kDa between the positions of HC and CI (Domier *et al.*, 1986). Putative proteins of 29-42 kDa have been proposed for other potyviruses. Significant sequence homology has not been reported in these regions between the few potyviruses for which information is available or between these putative proteins and others in protein sequence data banks.

Lain *et al.*, (1989) have drawn attention to a previously noted potential cleavage site in the amino acid sequence between HC and CI which could generate a small protein of approximately 6 kDa.

Cylindrical Inclusion Protein (CI)

There is an extensive literature dealing with taxonomic, structural and biochemical considerations of potyviral cylindrical inclusions (recently reviewed by Lesemann, 1988). These inclusions, which are found in the cytoplasm of cells infected with any of the potyviruses, are of rather exotic shape (Edwardson, 1974) and are thought to be composed entirely of a single protein (CI) of about 70 kDa which is encoded by the viral RNA (Dougherty and Hiebert, 1980a).

The amino acid sequences of the CI regions of the polyproteins of TVMV, TEV, PVY and PPV are quite similar, showing 53-60% sequence identity. Of particular note is the highly conserved "NTP motif" sequence Gly-X-X-Gly-X-Gly-Lys-Ser (where X = any amino acid residue) which is similar to the nucleotide binding site characteristic of ATP and GTP binding proteins and which is also present in non-structural viral proteins thought to be involved in replication of the RNAs of many viruses (Gorbalenya *et al.*, 1988; Hodgman, 1988; Zimmern, 1988). Lain *et al.*, (1989) report the existence of several regions of amino acid sequence homology between potyviral CIs and several helicases and raise the possibility of a role of the CI in unwinding of double-stranded replicative forms during synthesis of progeny viral RNA (Gorbalenya *et al.*, 1988). As noted below, these regions of sequence homology and the similarity in location and order of certain genes between the potyviruses, picornaviruses and comoviruses (Domier *et al.*, 1987; Goldbach, 1987) support the notion of a role of the CI in replication.

It has been proposed by Langenberg (1986) that cylindrical inclusions may be involved in the intercellular movement of potyvirus particles through plasmodesmata. Whether the inclusions might function in aligning virus particles so that they are in position to be transported or are more directly involved in the passage of the particles through the plasmodesmata, or whether some other mechanism might be involved, was not settled.

"6 kDa Protein"

On the basis of the locations in the TVMV, TEV, PVY and PPV polyproteins of apparent consensus protease cleavage sites, there may be a gene for a polypeptide of approximately 6 kDa between the CI and NIa genes. At first glance, it would seem that this could be the VPg gene since TEV RNA was reported years ago to contain a covalently-linked protein of approximately 6 kDa (Hari, 1981). In addition, the position in the genome of such a gene would be consistent with those of the comoviral and picornaviral VPgs (Kitamura *et al.*, 1981; Zabel *et al.*, 1984). However, more recent investigations provide evidence that the VPgs of TVMV, PPV and TEV are much larger than 6 kDa (Siaw *et al.*, 1985; Shahabuddin *et al.*, 1988; Reichmann *et al.*, 1989; Murphy *et al.*, 1989). This leaves in question the function of a 6 kDa polypeptide if, indeed, it is processed from the polyprotein at some stage during infection.

NIa or 49K Proteinase

Potyvirus-infected cells contain two virus-encoded proteins which in a limited number of cases are present in aggregates known as nuclear inclusions (Edwardson, 1974; Knuhtsen *et al.*, 1974). Why inclusion formation is not a consistent feature of the group and why such inclusions should be located in the nuclei of infected cells are two quite interesting but unanswered questions.

The smaller of the two nuclear inclusion proteins has been referred to as the 49K proteinase in the case of TEV and the NIa with TVMV. This protein is a viral-encoded protease which performs several cleavages of the potyviral polyprotein (Carrington and Dougherty, 1987; Hellmann *et al.*, 1988; Dougherty and Carrington, 1988). The series of elegant experiments in which the proteolytic activity of the enzyme and the specific polyprotein cleavage sites of TEV have been characterized and defined are described in the following chapter.

The VPg of TVMV has been reported to be a protein of 24 kDa in size (Siaw *et al.*, 1985) and to be co-amino-terminal with NIa (Shahabuddin *et al.*, 1988). More recently, it has been shown that purified TEV RNA contains bound protein molecules of 24 and 49 kDa and that both of these proteins react with anti-TEV 49K serum (Murphy *et al.*, 1989). In none of these investigations was a protein of 6 kDa detected in highly-purified preparations of viral RNA. These reports suggest that the amino-terminal half of or the entire small nuclear inclusion protein is the potyviral VPg. How the amino-terminal half is cleaved from the rest of the protein, or why with some molecules it is not, is not known. The amino acid sequence of the protein does not reveal an obvious site at which such a proteolytic cleavage event might occur.

The function of the potyviral VPg has not been established but by analogy with poliovirus, it seems likely that it plays a role in the initiation of replication (Wimmer, 1982) and thus forms the 5' terminus of the progeny RNA. With increasing evidence of a bi-functional role for the potyviral NIa or 49K proteinase, it might be useful to consider referring to it as the VPg-PRO protein.

NIb or 58K Protein

The larger of the two potyviral nuclear inclusion proteins has been much less investigated than the smaller. There are, however, persuasive reasons to believe that it functions as an RNA-dependent RNA polymerase, or a subunit thereof, in the synthesis of progeny viral RNA. Two regions in the TEV 58K and TVMV NIb proteins were found (Allison *et al.*, 1986; Domier *et al.*, 1987) to display significant sequence homology with regions in other proteins known or thought to be involved in nucleic acid polymerase reactions (Kamer and Argos, 1984; Argos, 1988), and this feature has remained consistent as additional potyviral nucleotide sequence information has become available. These blocks of sequence are Ser-Gly-Gln-Pro-Ser-Thr-Val-Val-Asp-Asn-Thr/Ser (amino acid residues 310-320 from the putative amino-terminus of TVMV NIb; residues 2534-2544 in the polyprotein) and, some 31 residues downstream in TVMV NIb, Gly-Asp-Asp residues bordered on both sides by hydrophobic residues. It may also be pertinent to note that a 19-amino acid region nearer the putative amino-termini of some potyviral NIb proteins has been shown to have significant homology with a region in a putative coronaviral polymerase (Gorbalenya *et al.*, 1989).

Of particular interest is the similarity of the potyviral NIb proteins to the putative "core" polymerases in the 3D protein of poliovirus and the 87 kDa protein encoded by CPMV B-RNA. In fact, the entire set of four proteins, CI-VPg-NIa-NIb, is equivalent in gene order as well as regions of amino acid homology to the 58 kDa-VPg-24 kDa-87 kDa arrangement in B-RNA of CPMV and to that of the 2C-VPg-3C-3D proteins in the picornaviruses (Domier *et al.*, 1987; Goldbach, 1987). These similarities (Fig. 3) provide attractive evidence that this group of potyviral proteins is involved in a membrane bound complex in which replication of viral RNA occurs and prompted Goldbach (1987) to place the potyviruses in the picornaviral "supergroup" of viruses.

Among the potyviruses themselves, it is the NIb protein, with 55-63% sequence identity, that is the most conserved of all the viral proteins. This would seem a logical situation for a viral protein suspected of being involved in what, among a group of related but distinct viruses, is perhaps the least unique activity.

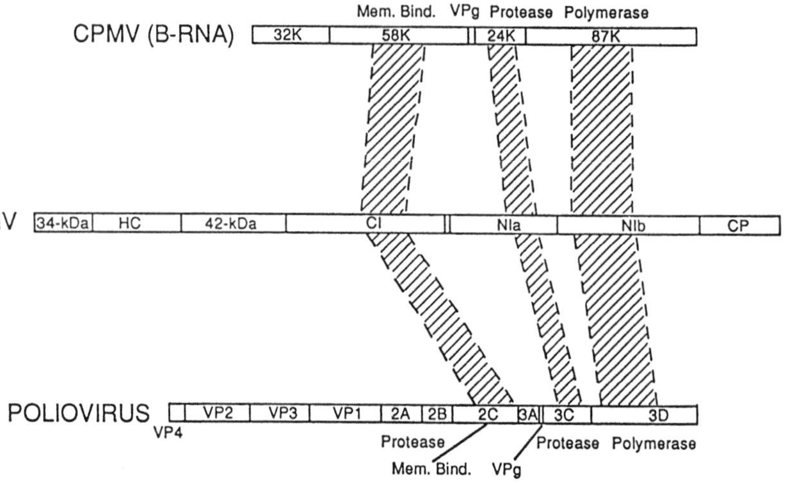

Fig. 3 Comparison of gene maps of TVMV RNA, CPMV B-RNA and poliovirus RNA. Shaded areas point to regions of homology identified by comparisons of predicted amino acid sequences. Modification of figure in Domier *et al.*, (1987).

Coat Protein

The coat proteins of potyviruses vary in size from 30-37 kDa and are arranged in the virion such that their amino- and carboxy-termini are located on or near the surface of the particle (Dougherty *et al.*, 1985; Allison *et al.*, 1985; Shukla *et al.*, 1988). The amino acid sequences of quite a number of potyviral coat proteins have been determined and it appears that the sequence of the amino-terminal regions of different potyviruses is quite variable while the central and carboxy-terminal regions are highly homologous (Shukla and Ward, 1989).

The major role of the coat protein is, of course, its participation in the structure of the virion. Lack of, or poor, aphid transmissibility of certain isolates of TEV (Pirone and Thornbury, 1983), ZYMV (Antignus *et al.*, 1989) and turnip mosaic virus and TVMV (T.P. Pirone and D.W. Thornbury, unpublished data) can occur even when biologically active HC is produced in plants infected with these isolates. This suggests that the coat protein may also be involved in regulating transmission of virus by aphids.

Differences in amino acid sequence between the coat proteins of aphid-transmitted and non-transmitted strains of TEV indicate that the amino-terminal portions of these proteins may be involved in the natural spread of virus particles from plant to plant by aphids (Allison *et al.*, 1985). An Asp-

Ala-Gly motif (located at amino acid residue positions 5-7 in the TVMV and TEV coat proteins; position 2745-2747 in the TVMV polyprotein) is present in the coat proteins of several isolates of potyviruses which are transmitted by aphids but missing from some which are not (Harrison & Robinson, 1988). Analysis of the appropriate sequences in additional isolates is needed to verify this observation.

EXPRESSION OF POTYVIRAL GENES IN TRANSGENIC PLANTS

There has recently been much interest in and effort given to transforming plants with genes which encode structural and non-structural viral proteins. There are several reasons for this enthusiasm. The prospects of engineering crops for enhanced resistance to virus infection are especially promising (Powell Abel *et al.*, 1986). In addition, transgenic plant models seem likely to augment our understanding of the role of virus-encoded genes and gene products in host-range determination. Plant lines that express viral genes and mutant derivatives thereof will also be important tools in the genetic dissection of potyviral genomes when used in conjunction with *in vitro*-generated viral RNAs.

There are already a few reports of potyviral gene expression in transgenic plants. Cloned DNA representing all but the 3'-terminal 24 nucleotides of the TVMV CI gene has been used to transform tobacco plants (Graybosch *et al.*, 1989). The appropriate RNA and protein were detected in the plants but cylindrical inclusions were not.

Tobacco plants which have been transformed with cloned DNA containing the 34K-HC-42K genes of TVMV have recently been shown to express biologically active HC (Berger *et al.*, 1989). The protein is of the same size as HC isolated from infected, non-transgenic plants and is therefore being properly processed from the truncated polyprotein. Analysis of plants transformed with cloned DNA in which specific changes have been made may be useful in determining amino acid sequences essential to HC activity.

Transgenic tobacco and potato plants which express potyviral coat proteins and their reactions to subsequent inoculation with various potyviruses are described in the chapter by R.N. Beachy.

IN VITRO PRODUCTION OF INFECTIOUS POTYVIRAL RNA –TOWARD A GENETIC ANALYSIS OF POTYVIRUSES

To date, most of what we know of the molecular biology of the potyviruses comes from the use of *in vitro* experimental systems and cytopathological examinations. The results of these types of investigation have made it possible to describe some of the characteristics of the transmission of virus particles by aphids, the development and structure of inclusions and the molecular determinants of proteolytic processing. However, there are also many other vital reactions which occur during the course of infection and disease development and how these activities are expressed and regulated has yet to be understood. To this end, a genetic study of potyviruses seems one of the next logical steps.

The production of TVMV RNA *in vitro* from bacterial transcription vectors has recently been reported (Domier *et al.*, 1989). The RNA has been shown to generate infections with a low but useful efficiency in tobacco plants and protoplasts. This development makes feasible a detailed investigation of the functions and interrelationships of the various TVMV gene products. Although the possibility of extensive RNA recombination may hinder these efforts, it should be possible to derive a complementation map of the virus through the use of site-directed and random mutagenesis, mixed infections and transgenic plant lines that express various viral genes.

ACKNOWLEDGEMENTS

We wish to express our deep admiration of and gratitude to the many colleagues who have contributed to the potyvirus research which has been conducted at this institution. We also thank the U.S. Department of Agriculture, the Kentucky Tobacco and Health Research Institute and NATO for support of our efforts.

REFERENCES

Allison, R., Johnston, R.E., and Dougherty W.D. (1986). The nucleotide sequence of the coding region of tobacco etch virus genomic RNA: evidence for the synthesis of a single polyprotein. *Virology* **154**, 9-20.

Allison, R.F., Dougherty, W.G., Parks, T.D., Willis L., Johnston, R.E., Kelly, M. and Armstrong, F.B. (1985). Biochemical analysis of the capsid protein gene and capsid protein of tobacco etch virus: N-terminal amino acids are located on the virion's surface. *Virology* **147,** 309-316.

Antignus, Y., Raccah B., Gal-On, A., and Cohen, S. (1989). Biological and serological characterization of zuc-chini yellow mosaic virus and watermelon mosaic virus 2 isolate in Israel. *Phytoparasitica*, in press.

Argos, P. (1988). A sequence motif in many polymerases. *Nucleic Acids Res.* **16**, 9909-9916.

Berger, P.H., and Pirone, T.P. (1986). The effect of helper component on the uptake and localization of potyviruses in Myzus persicae. *Virology* **153,** 256-261.

Berger, P.H., Hunt A.G., Domier L.L., Hellmann, G.M., Stram Y., Thornbury, D.W. and Pirone, T.P. (1989). Expression in transgenic plants of a viral gene product that mediates insect transmission of potyviruses. *Proc. Natl. Acad. Sci. USA* in press.

Carrington, J.C., and Dougherty W.G. (1987). Small nuclear inclusion protein encoded by a plant potyvirus genome is a protease. *J. Virol.* **61,** 2540-2548.

Carrington J.C., Cary S.M., Parks, T.D., and Dougherty, W.G. (1989a). A second proteinase encoded by a plant potyvirus genome. *EMBO J.* **8,** 365-370.

Carrington, J.C., Freed, D.D., and Sanders T.S. (1989b). Autoproteolytic processing of potyvirus proteinase HC-Pro in Escherichia coli and *in vitro*. *J. Virol.* in press.

De Mejia, M.V.G., Hiebert, E., Purcifull D.E., Thornbury D.W., and Pirone, T.P. (1985). Identification of potyviral amorphous inclusion protein as a nonstructural, virus-specific protein related to helper component. *Virology* **142**, 34-43.

Deom, C.M., Oliver, M.J., and Beachy, R.N. (1987). The 30-kilodalton gene product of tobacco mosaic virus potentiates virus movement. *Science* **237**, 389-393.

Domier, L.L., Franklin, K.F., Shahabuddin M., Hellmann, G.M., Overmeyer, J.H., Hiremath, S.T., Siaw, M.F.E., Lomonossoff, G.P., Shaw, J.G. and Rhoads, R.E. (1986). The nucleotide sequence of tobacco vein mottling virus RNA. *Nucleic Acids Res.* **14**, 5417-5430.

Domier, L.L., Franklin K.M., Hunt, A.G., Rhoads, R.E., and Shaw, J.G. (1989). Infectious *in vitro* transcripts from cloned cDNA of a potyvirus, tobacco vein mottling virus. *Proc Natl. Acad .Sci. USA* **86**, 3509-3513.

Dougherty, W. G., and Hiebert, E. (1980). Translation of potyviral RNA in a rabbit reticulocyte lysate: identification of nuclear inclusion proteins as products of tobacco etch virus RNA translation and cylindrical inclusion protein as a product of the potyvirus genome. *Virology* **104**, 174-182.

Dougherty, W.G., Willis, L., and Johnston, R.E. (1985). Topographic analysis of tobacco etch virus capsid protein epitopes. *Virology* **144**, 66-72.

Dougherty, W.G., and Carrington, J.C. (1988). Expression and function of potyviral gene products. *Ann. Rev. Phytopath.* **26**, 123-143.

Dougherty, W.G., and Hiebert, E. (1980). Translation of potyvirus RNA in a rabbit reticulocyte lysate: cell-free translation strategy and a genetic map of the potyviral genome. *Virology* **104**, 183-194.

Edwardson, J.R. (1974). Some properties of the potato virus Y group. *Fla. Agric. Exp. Sta. Monogr. Ser.* **4** , 1-398.

Goldbach, R. (1987). Genome similarities between plant and animal RNA viruses. *Microbiol. Sci.* **4**, 197-202.

Gorbalenya, A.E., Koonin E.V., Donchenko, A.P., and Blinov, V.M. (1988). A conserved NTP-motif in putative helicases. *Nature* **333**, 22-22.

Gorbalenya, A.E., Koonin, E.V., Donchenko, A.P., and Blinov, V.M. (1989). Coronavirus genome: Prediction of putative functional domains in the non-structural polyprotein by comparative amino acid sequence analysis. *Nucleic Acids Res.* **17**, 4847-4861.

Govier, D.A., and Kassanis, B. (1971). The role of helper virus in aphid transmission of potato aucuba mosaic virus and potato virus C. *J. Gen. Virol.* **13**, 221-228.

Graybosch, R., Hellmann, G.M., Shaw, J.G., Rhoads, R.E., and Hunt, A.G. (1989). Expression of a potyvirus non-structural protein in transgenic tobacco. *Biochem .Biophys .Res. Commun.* **160**, 425-432.

Hammond, J., and Hammond, R.W. (1989). Molecular cloning, sequencing and expression in Escherichia coli of the bean yellow mosaic virus coat protein gene. *J. Gen. Virol.* **70**, 1961-1974.

Hari, V., Siegel, A., Rozek, C., and Timberlake, W.E. (1979). The RNA of tobacco etch virus contains poly(A). *Virology* **92**, 568-571.

Hari, V. (1981). The RNA of tobacco etch virus: further characterization and detection of protein linked to RNA. *Virology* **112**, 391-399.

Harrison, B.D., and Robinson, D.J. (1988). Molecular variation in vector-borne plant viruses: epidemiological significance. *Phil. Trans. Roy. Soc. London* **B321**, 447-462.

Hellmann, G.M., Shaw, J.G., Lesnaw, J..A., Chu, L.-Y., Pirone, T.P., and Rhoads, R.E. (1980). Cell-free translation of tobacco vein mottling virus RNA. *Virology* **106**, 207-216.

Hellmann, G.M., Hiremath, S.T., Shaw, J.G., and Rhoads, R.E. (1986). Cistron mapping of tobacco vein mottling virus. *Virology* 151, 159-171.encoded protease. *Virology* **163**, 554-562.

Hiebert, E., and Dougherty, W.G. (1988). Organization and expression of the viral genomes. In:" The Plant Viruses. Vol. 4. The Filamentous Plant Viruses" (R.G.Milne, Ed.) pp. 159-178. Plenum Press, New York, NY.

Hill, J.H., and Benner, H.I. (1976). Properties of potyvirus RNAs: turnip mosaic, tobacco etch and maize dwarf mosaic virus. *Virology* **75**, 419-432.

Hodgman, T.C. (1988). A new superfamily of replicative proteins. *Nature* **333**, 22-23.

Kamer, G., and Argos, P. (1984). Primary structural comparison of RNA-dependent polymerases from plant, animal and bacterial viruses. *Nucleic Acids Res.* **12**, 7269-7282.

Kitamura, N., Semler, B.L., Rothberg, P.G., Larsen, G.R., Adler, C.J., Dorner, A.J., Emini, E/A/. Hanecak, R., Lee, J.J., van der Werf, S., Anderson, C.W. and Wimmer, E. (1981). Primary structure, gene organization and polypeptide expression of poliovirus RNA. *Nature* **291**, 547-553.

Knuhtsen, H., Hiebert, E., and Purcifull, D.E. (1974). Partial purification and some properties of tobacco etch virus induced intranuclear inclusions. *Virology* **61**, 200-209.

Langenberg, W.G. (1986). Virus protein association with cylindrical inclusions of two viruses that infect wheat. *J. Gen. Virol.* **67**, 1161-1168.

Lain, S., Riechmann, J.L., and Garcia, J.A. (1989). The complete nucleotide sequence of plum pox potyvirus RNA. *Virus Res.* **13**, 157-172.

Lecoq, H. (1986). A poorly aphid transmitted variant of zucchini yellow mosaic virus. *Phytopathology* **76**, 1063.

Leonard, D.A., and Zaitlin, M. (1982). A temperature-sensitive strain of tobacco mosaic virus defective in cell-to-cell movement generates an altered viral-encoded protein. *Virology* **117**, 416-424.

Lesemann, D.-E. (1988). Cytopathology. In:" The Plant Viruses. Vol. 4. The Filamentous Plant Viruses" (R.G.Milne, Ed.) pp. 179-235. Plenum Press, New York, NY.

Maiss, E., Timpe, U., Brisske, A., Jelkmann, W., Casper, R., Himmler, G., Mattanovich, D. and Katinger, H.W.D. (1989). The complete nucleotide sequence of plum pox virus RNA. *J.Gen.Virol.* **70**, 513-524.

McDonald, J.G., and Bancroft, J.B. (1977). Assembly studies on potato virus Y and its coat protein. *J. Gen. Virol.* **35**, 251-263.

Milne, R.G. (1988). The economic impact of filamentous plant viruses. In:" The Plant Viruses. Vol. 4. The Filamentous Plant Viruses" (R.G.Milne, Ed.) pp. 331-335. Plenum Press, New York, NY.

Murant, A.F., Raccah, B., and Pirone, T.P. (1988). Transmission by vectors. In:" The Plant Viruses. Vol. 4. The Filamentous Plant Viruses" (R.G.Milne, Ed.) pp. 237-273. Plenum Press, New York, NY.

Murphy, J.F., Rhoads, R.E., Hunt, A.G., and Shaw, J.G. (1989). Tobacco etch virus RNA has a 24-kDa VPg. *Virology* in press.

Oh, C.-S., and Carrington, J.C. (1989). Identification of essential residues in potyvirus proteinase HC-Pro by site directed mutagenesis. *Virology* in press.

Pirone, T.P., and Thornbury, D.W. (1983). Role of virion and helper component in regulating aphid transmission of tobacco etch virus. *Phytopathology* **73**, 872-875.

Powell Abel, P., Nelson, R.S., Hoffmann, N., Rogers, S.G., Fraley, R.T., and Beachy, R.N. (1986). Delay of disease development in transgenic plants that express the tobacco mosaic virus coat protein gene. *Science* **232**, 738-743.

Reichmann, J.L., Lain, S., and Garcia, J.A. (1989). The genome-linked protein and 5' end RNA sequence of plum pox potyvirus. *J. Gen. Virol.* **70**, 2785-2789.

Robaglia, C., Durand-Tardif, M., Tronchet, M., Boudazin, G., Astier-Manifacier, S., and Cassedelbart, F. (1989). Nucleotide sequence of potato virus Y (N strain) genomic RNA. *J .Gen. Virol.* **70**, 935-947.

Sehnke, P.C., Mason, A.M., Hood, S.J., Lister, R.M., and Johnson, J.E. (1989). A "zinc-finger"-type binding domain in tobacco streak virus coat protein. *Virology* **168**, 48-56.

Shahabuddin, M., Shaw, J.G., and Rhoads, R.E. (1988). Mapping of the tobacco vein mottling virus VPg cistron. *Virology* **163**, 635-637.

Shukla, D.D., Strike, P.M., Tracy, S.L., Gough, K.H., and Ward, C.W. (1988). The N-terminus and C-terminus of the coat proteins of potyviruses are surface-located and the N-terminus contains the major virus-specific epitopes. *J. Gen. Virol.* **69**, 1497-1508.
Shukla, D.D., and Ward, C.W. (1989). Identification and classification of potyviruses on the basis of coat protein sequence data and serology. *Arch. Virol.* in press.
Siaw, M.F.E., Shahabuddin, M., Ballard, S., Shaw, J.G., and Rhoads, R.E. (1985). Identification of a protein covalently linked to the 5' terminus of tobacco vein mottling virus RNA. *Virology* **142**, 134-143.
Thornbury, D.W., Hellmann, G.M., Rhoads, R.E., and Pirone, T.P. (1985). Purification and characterization of potyvirus helper component. *Virology* **144**, 260-267.
Tollin, P., and Wilson, H.R. (1988). Particle structure. In:" The Plant Viruses. Vol. 4. The Filamentous Plant Viruses" (R.G.Milne, Ed.) pp. 51-83. Plenum Press, New York, NY.
Turpen, T. (1989). Molecular cloning of a potato virus Y genome: Nucleotide sequence homology in non-coding regions of potyviruses. *J. Gen. Virol.* **70**, 1951-1960.
Wimmer, E. (1982). Genome-linked proteins of viruses. *Cell* **28**, 199-201.
Zabel, P., Moerman, M., Lomonossoff, G., Shanks, M., and Beyreuther, K. (1984). Cowpea mosaic virus VPg: sequencing of radiochemically modified protein allows mapping of the gene on B RNA. *EMBO J.* **3**, 1629-1634.
Zimmern, D. (1988). Evolution of RNA viruses, vol. 2, Retrovirus, viroids and RNA recombination. In:" RNA Genetics" (Domingo, E.,, Holland, J.J., and Ahlquist, P., Eds.) pp. 211-240. CRC Press Inc, Boca Raton, FL.

DISCUSSION OF J. SHAW'S PRESENTATION

D. French: What symptoms do tobacco plants show that are doubly infected with TVMV and TEV?
J. Shaw: There is no noticeable synergism.

R. Allison: You mentioned that transgenic plants containing the CI-gene showed enhancement of symptoms upon inoculation with TEV. What were the symptoms when you inoculated with TVMV?
J. Shaw: We haven't done this experiment yet.

R. *Allison*: Why did you check TEV before you checked TVMV?
J. Shaw: We began with TEV because TVMV symptoms on young plants are very difficult to detect. The symptoms seem to go through a stage where you think you can see them and then they go away and finally show up again. With TEV there is a much more distinct and clear reaction.

W. Kaniewski: How many independent transgenic lines expressing the TVMV coat protein did you examine to draw the conclusion that such plants show tolerance for this virus?
A. Hunt: The effects on CI-transgenic plants were seen with two independent plant lines. Concerning the coat protein-transgenic plants, we isolated a lot of independent lines expressing low levels of coat protein, that had no

biological effects. The highest expressing line is the one showing the effects that have been discussed by Shaw.

J. Hammond: Is TVMV the only virus you tested against the coat protein transgenic plants?

J. Shaw: We have tested TEV and TVMV. With TEV we found tolerance, with TVMV we get no protective effect at all. That was the mysterious effect I was referring to.

R. Goldbach: I always thought the 6 K protein would represent the VPg but now you showed evidence that VPg would be a 24 K protein. What is the hard evidence that this 24 K protein is really the VPg? If you could prove that this 24 K protein is covalently linked to the RNA then I would be convinced. Can you exclude the possibility that the 24 K protein is a contaminant, a breakdown product of the NIa-protein, which has been co-purified, while on the other hand, VPg is a smaller protein as has been found for CPMV?

J. Shaw: This is entirely possible. We do not have proof of it. When we examined RNA that has been carried through several sucrose gradients after disruption of virus particles, we can indeed see some contaminating coat protein, but not a 6 K protein. When this RNA is then run through CsCl gradients we no longer see anything but the two proteins I have mentioned. The RNA has been quite exhaustively purified but just because we do not detect another protein does of course not say that it is not there. We are currently attempting to define the linkage between the VPg and the RNA.

N. Tumer: Did you introduce any sequence changes into the coat protein during the process of making the coat protein construct?

L. Domier: We added a methionine at the N-terminus. No other amino acid changes have been introduced.

P. Palukaitis: Are cylindrical inclusions formed in the transgenic plants expressing the CI-protein?

J. Shaw: No. Actually we didn't use a complete CI-construct, it is missing a few amino acids at the C-terminus.

R. Goldbach: You succeeded in producing infectious transcripts from a cDNA-clone and according to your PNAS article you prepared transcripts containing two extra G-residues at the 5'-terminus, and having low infectivity. In the case of CPMV, increase in infectivity was obtained when one of these two extra G-residues was deleted. Are you currently preparing transcripts with only one extra G or are you intending to trim the T7 promoter such that you get transcripts as natural as possible?

J. Shaw: Actually, one of the T7 constructs we used produced transcripts that had only one extra G-residue. So we are talking about two types of transcripts, having either one or two extra Gs. Both transcripts are as far as we know equally infectious.

Expression of the Potyvirus Genome: The Role of Proteolytic Processing

William G. Dougherty, T. Dawn Parks, Holly A. Smith and John A. Lindbo

Department of Microbiology, Oregon State University,
Corvallis, OR 97331, USA

The potato virus Y [potyvirus] group is comprised of a number of distinctive members, estimated to be between 100 and 150 dependent on the criteria used in classification (Edwardson, 1974; Hollings and Brunt, 1981). Although each member has a rather restrictive host range, the large number of putative members make these viruses ubiquitous pathogens present in most agricultural settings. As such, over the past 5 years they have become one of the more intensively studied plant virus groups. Using molecular biological approaches, our understanding of potyvirus gene structure, function, and expression has increased dramatically.

There are a number of characteristics shared by all members of the potyvirus group (for a recent review, Dougherty and Carrington, 1988). The genomic nucleic acid is a single-stranded RNA approximately 10,000 nucleotides in length. This RNA is encapsidated by numerous copies of a single type of capsid protein monomer with a molecular weight usually between 30,000 and 35,000 daltons (35kDa) when a number of different members are examined. The genomic RNA has a protein [VPg] instead of a 'cap structure' covalently attached to the 5' terminal nucleotide (Hari, 1981; Siaw *et al.*, 1985); and the 3' terminal region has a polyadenlyate region (Hari *et al.*, 1979; Allison *et al.*, 1984). Collectively, this RNA, the VPg molecule, and capsid protein monomers assemble into a flexuous rod-shaped particle usually 700-800 nm in length and 12-15 nm in diameter. A common characteristic of all potyvirus infections is the formation of cytoplasmic pinwheel-shaped inclusion bodies (Edwardson, 1974; Christie and Edwardson, 1977; Hiebert *et al.*, 1986). These bizarre-shaped structures are comprised of many copies of a single viral-encoded protein, which frequently has a molecular weight around 70,000. It is likely this protein is involved in replication, although a cell-movement function has been proposed for it also (Langenberg, 1986). In addition to the cytoplasmic pinwheel inclusion body, certain potyviruses accumulate other viral-encoded proteins into distinctive inclusion structures. An amorphous inclusion body has been described in and isolated from pepper mottle virus [PeMV]- and papaya ringspot virus-type W [PRV]-infected cells (de Mejia *et al.*, 1985); while nuclear inclusion bodies have been described and purified from tobacco etch virus [TEV]- (Knuhtsen *et al.*, 1974), clover yellow vein virus-, and bean yellow mosaic virus [BYMV]-infected plant tissue (Chang *et al.*, 1988). The amorphous inclusion body is an aggregate of an ~55kDa protein, while the nuclear inclusion body is a co-crystal formed by aggregation of two viral-encoded proteins of approximately 49kDa and 58kDa. Circumstantial evidence suggests that the amorphous inclusion protein is the 'helper factor protein' involved in aphid transmission, while the 49kDa and 58kDa nuclear inclusion proteins of TEV have been proposed to be a proteinase and RNA-dependent RNA-polymerase, respectively (Allison *et al.*, 1986).

A number of biochemical and molecular biological studies jointly suggest that the potyviral RNA genome is expressed into a large polyprotein which is co- and post-translationally processed by viral-encoded proteolytic activities. Animal picornaviruses and the plant como- and nepoviruses use a similar strategy. In fact, in all four virus groups, common themes in genome organization, cleavage sites, and proteinase characteristics are maintained (for an excellent review, see; Nicklin *et al.*, 1986; Wellink and vanKammen, 1988; or Krausslich and Wimmer, 1988). It is the intent of this chapter to summarize our current understanding of the proteolytic processing events mediated by one of the proteinases encoded by the potyvirus genome; namely the 49kDa protein of TEV and its analogue in other potyviruses. We will discuss features of the proteinase derived from our biochemical and molecular biological studies and summarize our current understanding of a cleavage site. Surprising results have emerged from these latter studies which implicate this potyvirus proteinase as something quite unique in nature.

POTYVIRUS GENOMIC ORGANIZATION AND EXPRESSION

A number of reports have concluded that proteolytic processing of a high molecular weight polyprotein precursor is involved in potyvirus genome expression. Early studies with TEV correlated the synthesis of high molecular weight precursors and the lack of subgenomic mRNAs in infected plant tissue and suggested a polyprotein mode of gene expression (Dougherty, 1983). Subsequent analyses of soybean mosaic virus [SMV] RNA loaded onto ribosomes or SMV RNA purified from virions and translated in the presence of amino acid analogues supported such a mode of expression (Vance and Beachy, 1984a,b). Yeh and Gonsalves (1985) were then able to show differential proteolytic processing of a polyprotein depending on the presence or absence of dithiothreitol. Nucleotide and amino acid sequence data were then presented for the coat protein gene of TEV-NAT (not-aphid-transmitted) isolate (Allison *et al.*, 1984) and PeMV (Dougherty *et al.*, 1985), suggesting the involvement of proteolytic processing in capsid protein formation. A year later, the entire nucleotide sequences of tobacco vein mottling virus [TVMV] (Domier *et al.*, 1986) and TEV-HAT (highly-aphid-transmitted) isolate (Allison *et al.*, 1986) were reported, which demonstrated the existence of a single open-reading frame over 9000nt in length in both viral genomes. Recently, the sequence data base for potyviral genomes has been greatly expanded with complete nucleotide sequences reported for plum pox virus [PPV] (Maiss *et al.*, 1989) and potato virus Y [PVY] (Robaglia *et al.*, 1989). Additionally, partial nucleotide sequence is now known for BYMV, another isolate of PPV, SMV, PRV, watermelon mosaic virus-I [WMV-I], wheat streak mosaic virus [WSMV], zucchini yellow mosaic virus [ZYMV],and sugarcane mosaic virus [SCMV]. Collectively, the sequence data and immunoprecipitation analyses of potyvirus gene products support a gene order and a mode of gene expression likely to be similar to that presented for TEV in Fig. 1. Ribosomes attach near the 5' terminus and begin translation at an AUG codon located at nucleotide position

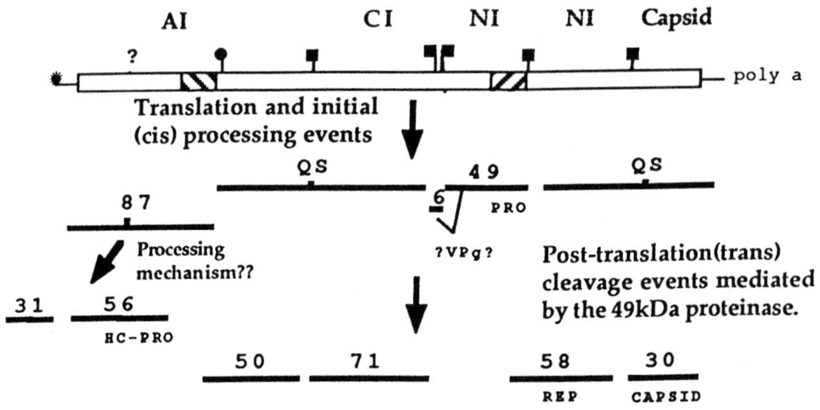

FIGURE 1. Schematic representation of the tobacco etch virus (TEV) genome and postulated mode of gene expression. The TEV genome is 9496 nucleotides in length and has a VPg at the 5' terminus and a polyadenylate region at the 3' terminus. The protein product associated with inclusion bodies is indicated above the line [AI = amorphous inclusion; CI = cytoplasmic pinwheel-shaped inclusion; NI = nuclear inclusion]. The capsid protein is also indicated. Translation of the TEV genome results in the synthesis of a large polyprotein (~346kDa) which is processed by two viral encoded activities. The 87kDa protein is formed by an autocatalytic cleavage at a Gly-Gly dipeptide which is mediated by a cysteine-like proteinase activity. The 49kDa and 6kDa proteins are released in an autocatalytic fashion mediated by a serine-like proteinase activity mapped to the C-terminal half of the 49kDa protein. Cleavage is between a Gln-Ser or a Gln-Gly. This proteinase is also responsible for the bi-molecular cleavage events at the 50/71kDa and the 58/30kDa junctions. The putative replicase (REP) function, helper factor-proteinase (HC-PRO) and proposed VPg molecules are labeled. Abbreviations used in this figure are: Gly-Gly = ⚲ ; Gln-Ser or Gln- Gly ▀ ; VPg = * ; Cysteine-like ▧ or serine-like ▨ proteolytic activities.

145. Soon after the 87kDa protein is synthesized, it autocatalytically releases from the rest of the growing polyprotein by cleavage between a Gly-Gly dipeptide (Carrington *et al.*, 1989a). This is an obligatory *cis* reaction and the C-terminal ~150 (but not 110) amino acids of the 87kDa protein contain a putative cysteine-like proteolytic activity inferred from mutagenesis studies (Carrington *et al.*, 1989a,b; Oh and Carrington, 1989). The rest of the growing polyprotein chain continues to elongate and, soon after the synthesis of the 49kDa protein, this amino acid sequence autocatalytically releases from the

polyprotein by cleaving at Gln-Gly dipeptides located at the N- and C-termini of the 49kDa protein. The 49kDa proteinase also rapidly cleaves the polyprotein at an upstream Gln-Ser dipeptide to release a small 6kDa protein (Carrington and Dougherty, 1987a, b). The 49kDa proteinase has been shown to cleave, in a bi-molecular or *trans* reaction, the Gln-Ser dipeptide between the 58kDa/30kDa protein junction and at the Gln-Ser dipeptide at the 50kDa/71kDa product junction (Carrington *et al.*, 1988). Attempts to demonstrate other cleavage events mediated by the 49kDa or 87kDa-related proteolytic activities in cell-free assays have been unsuccessful, and it is unclear how the 'putative' 87kDa polyprotein is processed to the 56kDa protein product observed in infected plant tissue. Therefore, in cell-free analyses, the 346kDa TEV polyprotein is processed to seven protein products by the catalytic activity of two viral-encoded proteinases.

PROTEINASE ACTIVITY ASSOCIATED WITH THE TEV 49KDA NUCLEAR INCLUSION PROTEIN

The TEV 49kDa nuclear inclusion protein has proteolytic activity associated with it when isolated from plant cells as a component of the nuclear inclusion body. This proteolytic activity can also be synthesized in cell-free transcription/translation systems using TEV cDNA containing 49kDa protein coding sequences (Carrington and Dougherty, 1987a). Using this latter approach, a similar proteolytic activity has been demonstrated for TVMV (Hellmann *et al.*, 1988) and PPV proteins (Garcia *et al.*, 1989a).

A number of molecular genetic, biochemical, and immunological studies have been conducted which suggest that the TEV 49kDa proteinase is structurally and mechanistically analogous to a trypsin-like serine proteinase with two unique characteristics. First, the nucleophilic serine residue of the catalytic triad has been replaced with a cysteine residue in the TEV 49kDa proteinase. A similar replacement is proposed for picorna-, como-, and nepovirus proteinases. Secondly, this proteinase displays a unique degree of specificity towards its substrate; the domain(s) responsible for this substrate specificity have not been identified. In Fig. 2, the TEV 49kDa proteinase amino acid sequence is aligned with the published sequences of three other potyviral 49kDa- equivalent sequences and the amino acid sequence of the poliovirus 3C- and the CPMV 24kDa-proteinases. Key residues hypothesized to be involved in catalysis are emphasized.

The translation of TEV RNA isolated from virions or transcribed from TEV cDNA (or TVMV or PPV cDNAs) sequences containing the 49kDa coding sequence consistently results in the formation of a series of related proteins with molecular weights around 49kDa. In cell-free studies, this 49kDa protein arises even if transcribed segments of TEV cDNA are used that encode a much larger part of the polyprotein. Subsequent deletion and site-directed mutagenesis analyses have revealed a proteolytic domain contained in the C-terminal half of the 49kDa protein. However, only in TEV [and a few other

```
 TEV    1 G KKNQ KHKLKMREARGARGQYEVAAEPEALEHYFGSAYNNKGKRKGTTRGMGAKSRKFINMYGFDPTDFSYIRFVDPLTGHTIDESTNAPIDLVQHE
TVMV    1 G K   SRRRLQFRKARDDKMGYIMHGEGDTIEHFFGAAYTKKGKSKGKTHGAGTKAHKFVNMYGVSPDEYSYVRYLDPVTGATLDESPMTDLNIVQEH
 PPV    1 GFNRRQR QKLKFRQARDNRMAREVYGDDSTMEAYFGSAYSKKGKSKGKTRGMGTKTRKFVNMYGYDPTDYNFVRFVDPLTGHTLDESPLMDINLVQEH
 PVY    1 GKNKSKRIQALKFRHARDKRAGFEIDNNDDTIEEFFGSAYRKKGKGKGTTVGMGKSSRRFINMYGFDPTEYSFIQFVDPLTGRQIEENVYADIRDIQER

con       G--k-qr-qkLkfR-ARd-rmgyev-geddtiEh-FGsAY-kKGKsKG-TrGmGtksrkF-NMYGfdPt-ys--rfvDPlTGht-dEsp-adinlvQeh

 TEV   98 FGKVRTRMLIDDEIEPQSLSTHTTIHAYLVNSGTKKVLKVDLTPHSSLRASEKSTAIMGFPERENELRQTGMAVPVAYDQLPPKN E D  LTFEGESL
TVMV   96 FGEIRREAILADAMSPQ  QRNKGIQAYFVRNSTMPILKVDLTPHIPLKVCE SNNIAGFPEREGELRRTGPTETLPFDALPP    EKQEVAFESKAL
 PPV   99 FSQIRNDYIGDDKITMQHIMSNPGIVAYYIKDATQKALKVDLTPHNPLRVCDKTATIAGFPEREFELRQTGHPVFVEPNAIPKINEEGDEEVDHESKSL
 PVY  100 FSEVRKKMVENDDIEMQALGSNTTIHAYFRKDWCDKALKIDLMPHNPLKVCDKTNGIAKFPERELELRQTG P AVE  VDVK  DIPAQEVEHEAKSL

con       F-e-R--mi-dD-ie-Q-l-snt-IhAYfvkd-t-kaLKvDLtPHnpL-vc-k-n-IagFPERE-ELRqTG-pv-ve-dalp--n-e-dqev--EsksL

CPMV    1                                                                          Q/SS     VAIMSKCR A N  LVFGGT
PV      1                                                                          Q/GPG   FDYAVAMAK R N  IVTATTS

TEV   193 FKGPRDYNPISSTICHLTNESDGHTTSLYGIGFGPFIITNKHLFRRNNGTLLVQSL      HGVFKVKNTTTLQQHLID      GRDMIIIRMPK
TVMV  188 LKGVRDFNPISACVWLLENSSDGHSERLFGIGFGPYIIANQHLFRRNNGELTIKTM      HGEFKVKNSTQLQMKPVE      GRDIIVIKMAK
PPV   198 FRGLRDYNPIASSICQLNNSSGARQSEMFGLGFGGLIVTNQHLFKRNDGELTIRSH      HGEFVVKDTKTLKLLPCK      GRDIVIIRLPK
PVY   193 MRGLRDFNPIAQTVCRLKVSVEYGASEMYGFGFGAYIVANHHLFRSYNGSMEVQSM      HGTFRVKNLHSLSVLPIK      GRDIILIKMPK

con       f-GlRD-NPI-st-c-L-nssdgh-se--GiGFGpyI--NqHLFrrnnGelt-qsm      HGeFkVKntttLq-lpik      GRDiiiI-mpK

CPMV   19  NLQIVMVPGRR  FLACKHFFTHIKTKLRVEIVMDGRRYYHQFDPANIY   DIP      DSELVLYSHPSLEDVSHS      CWDLFCWDPDK
PV     22  KGEFTMLGVH                       DNVAILPTHAS   PGESIVIDGKEVEILDAKALEDQAGTNLEITIITLKRNEKFRDIRQHIPTQ
                                                 ▲                                               ▲
TEV   278 DFPPFPQKLKFREPQREERICLVTT    NFQTKSMSSMVSDTSCTFPSSDGIFWKHW IQTKDGQCGSPLVSTRDG  FIVGIHSASNFTNTNNYF
TVMV  273 DFPPFPQKLKFRQPTIKDRVCMVST    NFQQKSVSSLVSESSHIVHKEDTSFWQHW ITTKDGQCGSPLVSIIDG  NILGIHSLTHTTNGSNYF
PPV   283 DFPPFPRRLQFRTPTTEDRVCLIGS    NFQTKSISSTMSETSATYPVDNSHFWKHW ISTKDGHCGLPIVSTRDG  SILGLHSLANSTNTQNFY
PVY   278 DFPVFPQKLHFRAPTQNERICLVGT    NFQEKYASSIITETSTTYNIPGSTFWKHW IETDNGHCGLPVVSTADG  CIVGIHSLANNAHTTNYY

con       DFPpFPqkLkFR-Pt-e-R-Clvgt    NFQtKs-SS-vsetS-typ--ds-FWkHW I-TkdG-CG-PlVStrDG  -I-GiHSlan-tnt-Ny-

CPMV   99  ELPSVFGADFLSCKYNKFGGFYEAQYADIKVRTKKECLTIQSG   NYVNKVSRYLEYEAPTIPEDCGSLVIAHIGGKHKIVGVHVAGIQGKI
PV     95  ITETNDGVLIVNTSKYPNMYVPVG     AVTEQGYLNLGG       RQTARILMYNFPTRAGQCGGVITCTG    KVIGMHVGGNGSH
                                                                          ▲                *
TEV   368 TSVPKNF MELLTNQEAQQWVSGWRLNADSVLWGGHKVFMSKPEEPFQPVKEATQL MNELVYSQ
TVMV  363 VEFPEKF  VATYLDAADGWCKNWKFNADKISWGSFTLVEDAPEDDFMAKK TVAAIMDDLVRTQ
PPV   373 AAFPDNFETTYLSNQDNDNWIKQWRYNPDEVCWGSLQLKRDIPQSPFTICKLLTDL DGEFVYTQ
PVY   368 SAFDEDFESKYLRTNEHNEWVKSWVYNPDTVLWGPLKLKDSTPKGLFKTTKLVQDLIDHDVVVEQ

con       -afpenFe--yl-nqead-Wvk-WryN-D-vlWGslklk---Pe-pF---Kl-tdli---lVytQ

CPMV  193 GCASLLPPLEPIAQAQ/G
PV    168 GFAAALKRSYFTQSQ/G
```

Fig. 2 A comparision of the amino acid sequences of four potyviral 49kDa proteins and the poliovirus 3C (PV) and CPMV 24kDa(CPMV) proteinases. A consensus sequence is presented along with the three residues thought to play a role in catalysis (▲) and a residue proposed to be involved in substrate binding (✱). The four potyviruses are; tobacco etch virus (TEV), tobacco vein mottling virus (TVMV), plum pox virus (PPV), and potato virus Y (PVY).

potyviruses] can a similar protein be isolated from infected tissue (Dougherty and Carrington, 1988; Slade *et al.*, 1989). These analyses are presented in a schematic drawing in Fig. 3 and pertinent features of the proteinase can be summarized in the following statements.

1. The TEV 49kDa protein is 430 amino acids in length and is formed by obligatory autocatalytic [or *cis*] cleavage events which release the proteinase from the polyprotein substrate. Cleavage is between Gln-Gly dipeptides found at termini of the 49kDa proteinase (Carrington and Dougherty, 1987a, b). The 49kDa protein isolated as a component of the NI body has proteolytic activity.
2. Computer analysis and protein modeling suggest that the C-terminal half of these potyviral proteinases may be structurally similar to trypsin, a serine proteinase (Bazan and Fletterick, 1988; Gorbalenya *et al.*, 1989; Dougherty *et al.*, 1989b).

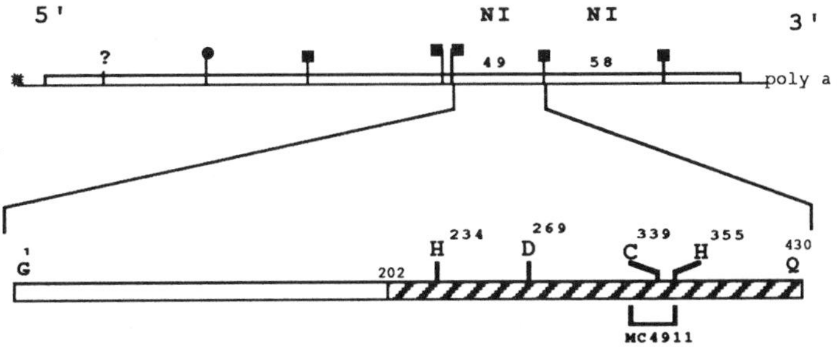

FIGURE 3. A schematic representation of pertinent aspects of the TEV 49kDa Protein
The TEV genome is presented at the top of the figure. Please see Figure 1 legend for an explanation of the symbols. The two proteins which constitute the nuclear inclusion body [NI] are indicated. An expanded schematic of the 49kDa proteinase is presented below the genomic map. The TEV 49kDa proteinase is 430 amino acids in length with the proteolytic domain localized to the C-terminal 228 amino acids [ ].The amino acid residues proposed to play a key role in catalysis or substrate binding are indicated along with their position in the 49kDa protein. The active site triad is proposed to be histidine, aspartic acid, and cysteine at TEV 49kDa amino acid positions 234, 269, and 339 respectively. The conserved histidine at position 355 is proposed to be involved in binding the P1 Gln of the substrate. The domain[aa318-352] which contains the epitope recognized by MC4911 is indicated.

3. Site-directed mutagenesis analyses of putative active site residues of the TEV 49kDa protein suggest the catalytic triad is comprised of His^{234}, Asp^{269} and Cys^{339}. The nucleophilic Cys^{339}, found in all proteinases encoded for by RNA viruses that use a polyprotein mode of gene expression, can be replaced with Ser, but the proteinase loses most (>99%) of its proteolytic activity. Other amino acid replacements at this TEV position result in a 49kDa proteinase which has no detectable activity (Dougherty *et al.*, 1989b).

4. Proteinase inhibitor studies suggest that the TEV 49kDa is similar to picorna- and como-viral encoded activities in that most inhibitors have no or only a marginal effect on the enzyme. The exceptions were Zn^{++} and general alkylating agents, such as iodoacetamide and N-ethylmaleimide, which eliminated activity (Dougherty *et al.*, 1989b).

5. A conserved His residue [in TEV, His^{355}] is always located 15-16 amino acids downstream of the active site Cys [in TEV, Cys^{339}] in RNA viral proteinases which are similar to the TEV 49kDa protein (i.e., picornaviral 3C proteinase, CPMV 24kDa proteinase). This residue has been implicated in binding the P1 Gln residue of the substrate (Bazan and Fletterick, 1988). Replacement of this residue eliminates cleavage, likely due to the inability of the substrate to bind to the proteinase (unpublished observations).

6. The 49kDa protein is likely a two-domain (function) protein. Deletion analysis suggests that the proteolytic function maps to the C-terminal half between amino acids 202 and 430 (Carrington and Dougherty, 1986). The analyses of the 49kDa equivalent of TVMV (Hellmann *et al.*, 1988) and PPV (Garcia *et al.*, 1989a) are consistent with this observation. Work from the group headed by John Shaw and Bob Rhoads with TVMV and TEV suggests that the N-terminal half is the genome-linked protein [VPg] (Shahabuddin *et al.*, 1988; Murphy *et al.*, 1989).

7. Proteolytic activity can be immuno-inhibited with polyclonal antiserum reactive with the TEV 49kDa nuclear inclusion protein, or with monoclonal antibody MC4911 (Slade *et al.*, 1989). This antibody has been mapped to an epitope which is located between TEV 49kDa amino acids 318 and 356. Other reactive monoclonal antibodies have no effect on proteolysis.

While our understanding of the TEV 49kDa proteinase has increased dramatically over the past 3 years, there remain a number of important issues to address. Namely, how is this proteinase able to achieve its unique degree of substrate specificity and, are this viral enzyme and the processes it mediates viable candidates for disruption in the design of antiviral strategies?

TEV 49KDA PROTEINASE CLEAVAGE SITE REQUIREMENTS

The TEV 49kDa proteinase recognizes an extended seven amino acid sequence at each naturally occurring cleavage site. It appears that other potyviral proteinases may recognize an extended amino acid sequence; however, the cleavage site sequences recognized by these proteinases are distinct, as shown in Table 1. In an interviral cell-free processing study, both TEV and TVMV proteinases and substrates have been synthesized; however, neither proteinase is able to process the alternate substrate to any detectable degree (Parks and Dougherty, unpublished observations). The TEV system is in contrast to poliovirus 3C proteinase, which cleaves at Gln/Ser or Gln/Gly dipeptides. The determining factor in picornaviral proteolytic processing at these dipeptides appears to be the accessibility of the dipeptide to the proteinase and the conformation of the 3C proteinase (Ypma-Wong *et al.*, 1988a,b).

The TEV 49K proteinase recognizes the consensus heptapeptide sequence:

P6		P5		P4		P3		P2		P1		P'1
Glu	-	Xaa	-	Xaa	-	Tyr	-	Xaa	-	Gln	-	Ser or Gly.

This amino acid sequence is found in only five locations on the TEV genome, and each is recognized and cleaved by the TEV 49kDa proteinase. cDNA sequences containing regions coding for these amino acids can be moved into other locations on the TEV genome and still maintain a functional cleavage site after proteins are produced. Additionally, oligonucleotide linkers containing DNA coding for only the seven amino acid sequence can be inserted into foreign locations, creating new functional cleavage sites when expressed. However, insertion of sequences coding for only a Gln/Ser dipeptide is not sufficient to direct cleavage (Carrington and Dougherty, 1988). A similar approach has been used with PPV with comparable results (Garcia *et al.*, 1989).

Of the seven amino acids comprising a TEV 49kDa proteinase cleavage site, four are conserved among naturally occurring sites in the TEV genome (Table 1). These are the Glu, Tyr, Gln, and Ser or Gly at positions P6, P3, P1 and P'1, respectively. Site-directed mutagenesis of each of these amino acids, using the

58K/30kDa cleavage site as a model, has shown each of these conserved residues is critical in promoting cleavage. Substantial reductions or an elimination of cleavage usually results when any of these amino acids are changed (Dougherty *et al.*, 1988). The remaining positions (P5, P4, and P2) in the heptapeptide sequence of a TEV 49kDa cleavage site have a variety of amino acids present when the five naturally occurring TEV cleavage sites are examined. Site-directed mutagenesis studies targeting these positions show a wide range of cleavage efficiencies with a variety of different amino acid replacements.

However, none of the substitutions at these positions totally eliminate cleavage. Additional replacement studies of amino acids located outside of the heptapeptide sequence (P7 and P'2) have a minimal effect on proteolytic cleavage, with two exceptions (Dougherty *et al.*, 1989a). The results of both the insertion studies and the mutagenesis studies lead to the conclusion that the TEV 49kDa proteinase recognizes a seven amino acid cleavage sequence. This sequence contains both conserved amino acids, which are essential for cleavage, and nonconserved amino acids, which can be substituted with varying effects on cleavage.

Differential processing of TEV cleavage sites

Based on the wide range of processing efficiencies observed in mutagenesis studies of the nonconserved positions, the hypothesis was developed that the amino acids found in the P5, P4, and P2 positions could influence the relative rate at which cleavage by the 49kDa proteinase proceeded. Initial cell-free studies to test this focused on two naturally occurring cleavage sites, the 50kDa/71kDa and the 58K/30kDa junctions. These two sites, both cleaved *in trans* by the 49kDa proteinase, differ in the amino acids found at the nonconserved P5, P4, and P2 positions (Fig. 1, 4, and Table 1). Following *in vitro* transcription and translation of cDNA sequences containing these two cleavage sites, 49kDa proteinase was added and processing was monitored over a 2 h time period, as shown in Fig. 4. In the case of the 50kDa/71kDa cleavage junction, processing proceeded such that one half of the precursor was processed to products approximately 25 min after the addition of proteinase. This was exhibited by conversion of the 32kDa precursor to 22kDa and 10kDa products, as detected by SDS-PAGE and subsequent autoradiography. Conversely, cleavage of the 58K/30K junction proceeded more rapidly, with half of the 34kDa precursor converted to 30kDa and 4kDa products 4.5 min after the addition of the 49kDa proteinase (Dougherty and Parks, 1989).

Site-directed mutagenesis of the two cDNA sequences containing these cleavage sites was performed to change the codons for the amino acids at the 50kDa/71kDa junction to those found at the 58kDa/30kDa site; and also to change the 58kDa/30kDa cleavage site sequence into that found at the 50kDa/71kDa junction (Fig. 4). Cell-free processing of these mutated substrates revealed that the 58kDa/30kDa heptapeptide cleavage sequence, when inserted into the context of the 50kDa/71kDa polyprotein, still processed at a

Table 1. Comparison of Known and Putative Potyviral Cleavage Sites

Virus[b]	Amino acid sequences at cleavage junctions[a,c]			Consensus	Citation
		P5 P3 P1	P'1		
TEV-HAT	50/70	-E-I-I-Y-T-Q	S		
	70/6	-E-T-I-Y-L-Q	S	P5 P3 P1 P'1	Allison *et al.*, 1986
	6/49	-E-P-V-Y-F-Q	S	**E**-X-X-**Y**-X-**Q** **S/G**	Carrington *et al.*, 1988
	49/58	-E-L-V-Y-F-Q	G		
	58/30	-E-N-L-Y-F-Q	S		
TEV-NAT	50/70	-E-I-I-Y-T-Q	S		
	70/6	-E-T-I-Y-L-Q	S		Allison *et al.*, 1985
	6/49	-E-P-V-Y-F-Q	G	**E**-X-X-**Y**-X-**Q** **S/G**	W.G. Dougherty,
	49/58	-E-L-V-Y-F-Q	G		unpublished
	58/30	-E-N-L-Y-F-Q	S		
TVMV	42/CI	-N-N-V-R-F-Q	S		
	CI/5.5	-E-A-V-R-F-Q	S	R F	Domier *et al.*, 1986
	5.5/NIa	-E-P-V-K-F-Q	G	-**V**-/-/-**Q** **S/G**	Hellmann *et al.*, 1988
	NIa/NIb	-D-L-V-R-T-Q	G	K T	
	NIb/Cap	-E-T-V-R-F-Q	S		
PPV-NAT	29/CI	-R-A-V-H-H-Q	S		
	CI/6	-E-C-V-H-H-Q	T	H	
	6/NIa	-E-E-V-I-H-Q	G	-**V**-X-/-**Q** X	Maiss *et al.*, 1989
	NIa/NIb	-E-F-V-Y-T-Q	S	T	
	NIb/Cap	-N-V-V-V-H-Q	A		
PVYn	42/CI	-Y-E-V-R-H-Q	S		
	CI/5.5	-Q-F-V-H-H-Q	A	H	
	5.5/NIa	-E-T-V-S-H-Q	G	-**V**-X-/-**Q** X	Robaglia *et al.*, 1989
	NIa/NIb	-D-V-V-V-E-Q	A	E	
	NIb/Cap	-Y-E-V-H-H-Q	A		
PPV	NIb/Cap	-N-V-V-V-H-Q	A		Lain *et al.*, 1988
PPV	NIb/Cap	-T-L-L-W-H-Q	A		Ravelonandro *et al.*, 1988
BYMV	NIb/Cap	-L-T-C-R-F-Q	S		Hammond and Hammond, 1989
PEMV	NIb/Cap	-Y-E-V-H-H-Q	A		Dougherty *et al.*, 1985
SCMV	NIb/Cap	-V-D-V-E-H-Q	S		Gough *et al.*, 1987
SMV	NIb/Cap	-E-S-V-S-L-Q	S		Eggenberger *et al.*,1989
WMV-1	NIb/Cap	-T-H-V-F-H-Q	S		Quemada *et al.*,unpubl.
PRV	NIb/Cap	-T-H-V-F-H-Q	S		Quemada *et al.*,unpubl.

[a]Abbreviations used: NI, nuclear inclusion; CI, cytoplasmic inclusion.
[b]The potyvirus sequences compared in this table are:tobacco etch virus-highly aphid transmissible isolate (TEV-HAT); and the TEV not-aphid transmissible isolate (TEV-NAT); tobacco vein mottling virus (TVMV); the plum pox virus not-aphid transmissible isolate (PPV-NAT); plum pox virus (PPV); bean yellow mosaic virus-GDD isolate (BYMV); pepper mottle virus (PeMV); potato virus Y-n isolate (PVYn); sugar cane mosaic virus-Johnson grass isolate (SCMV); soybean mosaic virus (SMV); watermelon mosaic virus (WMV-1); and papaya ringspot virus (PRV).
[c]Predicted cleavage sites are based on known processing sites of TEV and do not reflect those proposed by authors in all instances. For additional information pertaining to proposed viral products, please refer to original citations.

rate similar to that observed for the 58/30kDa polyprotein. The reciprocal experiment, with the 50kDa/71kDa heptapeptide cleavage sequence in the 58kDa/30kDa polyprotein background, showed that this sequence was processed at a 'slower' rate in this context as well. Replacements of individual amino acids at each of the nonconserved positions gave intermediate processing profiles, suggesting that all three positions played a role in defining the relative rate of cleavage (Dougherty and Parks, 1989).

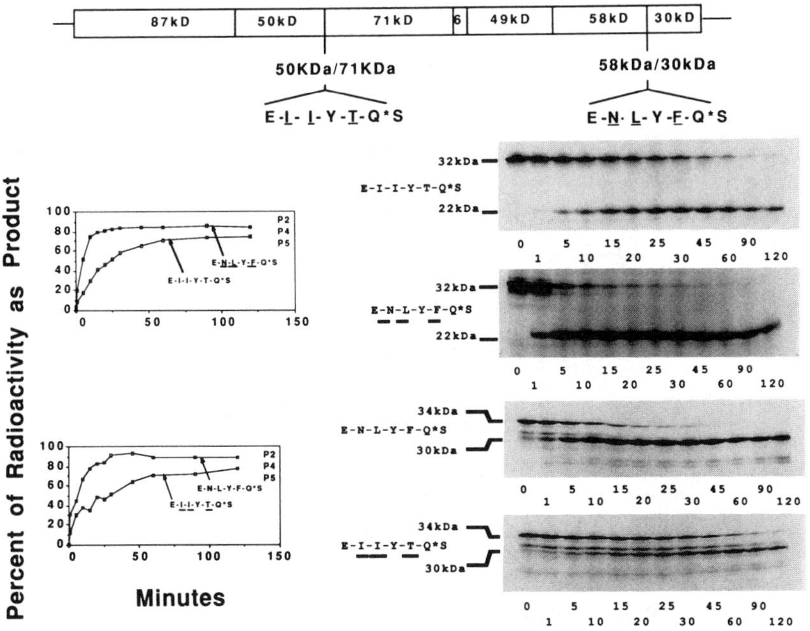

Fig. 4 Differential processing of two TEV 49kDa cleavage sites. A schematic of the TEV genome, with the amino acid sequences of the 50kDa/71kDa and 58kDa/30kDa cleavage sites, is shown at the top of the figure. After transcription and translation, substrates were incubated with TEV 49kDa proteinase and samples were removed at the time points indicated (min). After SDS-PAGE, the amount of radioactivity in precursor and product bands was determined. Quantitated results are shown in the graphs on the left as the percent of radioactivity found as processed product. The top graph shows the results of processing at the 50kDa/71kDa site using 32kDa substrates containing a naturally occurring site and a mutated variant containing the heptapeptide sequence found at the 58kDa/30kDa site. Autoradiographs of these reactions are shown at the right. The bottom graph shows the results of processing at the 58kDa/30kDa cleavage site, using 34kDa substrates containing the cleavage sequence found naturally at this site and a mutated substrate containing the 50kDa/71kDa cleavage site sequence. Autoradiographs of this set of the reactions are shown on the right. Amino acids are shown using the single letter code, with mutated amino acids underlined.

SUMMARY

The results from a number of different experimental approaches have permitted us to speculate that the amino acids found in the nonconserved positions may regulate the rate at which cleavage occurs at specific protein junctions on the TEV genome, and may represent a means of post-translational gene regulation. The polyprotein mode of gene expression prohibits TEV from utilizing transcription as a regulatory mechanism. One way that TEV could control relative amounts of gene products might be by differential processing of the polyprotein. In this scenario, the relative efficiency of a TEV 49kDa cleavage site could influence both the amount and the time of appearance of a mature gene product. A protein needed early in infection, or in larger quantities, could be located adjacent to a "rapid" site, while other proteins needed later could have "slow" sites. A precursor could have a function different from that of its cleaved products, with cleavage of a precursor converting an "early" gene product into a "late" gene product(s). A caution about these studies is that only cleavage events which are amenable to study in our 'defined' cell-free system have been examined to date. There may be other cleavage events which occur at relatively slow rates and are undetected in our assays. These cleavage events could be responsible for product formation or product inactivation and would be predicted to be essential in virus replication. Thus, the possibility exists that the TEV 49kDa proteinase is a major determinant of gene regulation during the TEV 'life cycle'.

ACKNOWLEDGEMENTS

We would like to thank the National Science Foundation and the Department of Energy for supporting our work. We are indebted to our colleagues who have assisted us in our studies (Jim Carrington, Dennis Hruby and crew, Susan Cary, Kristin Rorrer, Fernando Bazan and Bob Fletterick) and other scientists who have provided us with cDNA clones (John Shaw) and unpublished sequence data(J. Slightom, D. Gonsalves , H. Quemada, R. Beachy, S. Lommel, and J.Hammond).

REFERENCES

Allison, R.F., Sorenson, J.C., Kelly, M.E., Armstrong, F.B., and Dougherty, W.G. (1984). Sequence determination of the capsid protein gene and flanking regions of tobacco etch virus: evidence for the synthesis of a polyprotein in potyvirus genome expression. *Proc. Natl. Acad. Sci. USA* **82**: 3969-3972.

Allison, R.F., Johnston, R.E., and Dougherty, W.G. (1986). Nucleotide sequence of the coding region of tobacco etch virus genomic RNA: evidence for the synthesis of a single polyprotein. *Virology* **154**: 9-20.

Bazan, F.J., and Fletterick, R. (1988). Viral cysteine proteases are homologous to the trypsin-like family of serine proteases: structural and functional implications. *Proc. Natl. Acad. Sci. USA* **85**: 7872-7876.

Carrington, J.C., Cary, S.M., and Dougherty, W.G. (1988). Mutational analysis of tobacco etch virus polyprotein processing: *cis* and *trans* proteolytic activities of polyproteins containing the 49-kilodalton proteinases. *J. Virology* **62**, 2313-2320.

Carrington, J.C., Cary, S.M., Parks, T.D., and Dougherty, W.G. (1989a). A second proteinase encoded by a plant potyvirus genome. *EMBO J.* **8**, 365-370.

Carrington, J.C., and Dougherty, W.G. (1987a). Small nuclear inclusion protein encoded by a plant potyvirus genome is a protease. *J. Virology* **61**, 2540-2548.

Carrington, J.C., and Dougherty, W.G. (1987b). Processing of the tobacco etch virus 49K protease requires autoproteolysis. *Virology* **160**, 355-362.

Carrington, J.C., and Dougherty, W.G. (1988). A viral cleavage site cassette: identification of amino acid sequences required for tobacco etch virus polyprotein processing. *Proc. Natl. Acad. Sci. USA* **85**, 3391-3395.

Carrington, J.C., Freed, D.D., and Sanders, T.C. (1989b). Autocatalytic processing of the potyvirus helper component proteinase in Escherichia coli and *in vitro*. *J. of Virology* **63**: 4459-4463.

Chang, C.-A., Hiebert, E., and Purcifull, D.E. (1988). Purification, characterization, and immunological analysis of nuclear inclusions induced by bean yellow mosaic and clover yellow vein potyviruses. *Phytopathology* **78**: 1266-1275.

Christie, R.G., and Edwardson, J.R. (1977). Light and electron microscopy of plant virus inclusions. Fla Agric. Exp. Sta. Monogr. Ser. 9. 198pp.

De Mejia, M.V.G., Hiebert, E., and Purcifull, D.E. (1985). Isolation and partial characterization of the amorphous cytoplasmic inclusions associated with infections caused by two potyviruses. *Virology* **142**: 24-33.

Domier, L.L., Franklin, K.M., Shahabuddin, M., Hellmann, G.M., Overmyer, J.H., Hiremath, S.T., Siaw, M.F.E., Lomonossoff, G.P., Shaw, J.G., and Rhoades, R.E. (1986). The nucleotide sequence of tobacco vein mottling virus RNA. *Nucleic Acids Res.* **14**: 5417-5430.

Dougherty, W.G. (1983). Analysis of viral RNA isolated from tobacco leaf tissue infected with tobacco etch virus. *Virology* **131**, 473-481.

Dougherty, W.G., Allison, R.F., Parks, T.D., and Johnston, R.E. (1985). Nucleotide sequence at the 3' terminus of pepper mottle virus. *Virology* **146**: 282-291.

Dougherty, W.G., and Carrington, J.C., 1988. Expression and function of potyviral gene products. *Annual Review of Phytopathology* **26,** 123-143.

Dougherty, W.G., Carrington, J.C., Cary, S.M., and Parks, T.D. (1988). Biochemical and mutational analysis of a plant virus polyprotein cleavage site. *EMBO J.* **7,** 1281-1287.

Dougherty, W.G., Cary, S.M., and Parks, T.D. (1989a). Molecular genetic analysis of a plant virus polyprotein cleavage site: A model. *Virology* **171**: 356-364.

Dougherty, W.G., and Hiebert, E. (1980a). Translation of potyviral RNA in a rabbit reticulocyte lysate: identification of nuclear inclusion protein as products of tobacco etch virus RNA translation and cytoplasmic inclusion protein as a product of the potyvirus genome. *Virology* **104,** 174-182.

Dougherty, W.G., and Hiebert, E. (1980b). Translation of potyvirus RNA in a rabbit reticulocyte lysate: Cell-free translation strategy and a genetic map of the potyviral genome. *Virology* **104**: 183-194.

Dougherty, W.G., and Parks, T.D. (1989). Molecular genetic and biochemical evidence for the involvement of the heptapeptide cleavage sequence in determining the reaction profile at two tobacco etch virus cleavage site in cell-free assays. *Virology* **172**: 145-155.

Dougherty, W.G., Parks, T.D., Cary, S., Bazan, F.J., and Fletterick, R. J. (1989b). Characterization of the catalytic residues of the tobacco etch virus 49kDa proteinase. *Virology* **172**: 302-310.

Edwardson, J.R. (1974). Some properties of the potato virus Y-group. Fla. Agric. Exp. Sta. Monogr. Ser. 4. 398pp.

Eggenberger, A.L., Stark, D.M., and Beachy, R.N. (1989). The nucleotide sequence of a soybean virus coat protein-coding region and its expression in Escherichia coli, Agrobacterium tumefaciens, and tobacco callus. *J. Gen. Virology* **70**: 1853-1860.

Garcia, J.-A., Riechmann, J.L., and Lain, S. (1989a). Proteolytic activity of the plum pox potyvirus NIa-like protein in Escherichia coli. *Virology* **170**: 362-369.

Garcia, J.A., Reichmann, J.L., and Lain, S. (1989b) Artificial cleavage site recognized by plum pox potyvirus protease in Escherichia coli. *J. Virology* **63**: 2457-2460.

Gorbalenya, A.E., Donchenko, A.P., Blinov, V.M., and Koomin, E.V. (1989). Cysteine Proteases of positive strand RNA viruses and chymotrypsin-like serine protease; A distinct protein superfamily with a common structural fold. *FEBS Letters* **243**: 103-114.

Gough, K.,H., Azad, A.A., Hanna, P.J., and Shukla, D.D. (1987). Nucleotide sequence of the capsid protein and nuclear inclusion protein genes from the Johnson grass strain of sugarcane mosaic virus RNA. *J. Gen. Virology* **68**: 297-304.

Hammond, J., and Hammond, R. W.(1989). Molecular cloning, sequencing and expression in Escherichia coli of bean yellow mosaic virus coat protein gene. *J. Gen. Virology* **70**: 1961-1974.

Hari, V. (1981). The RNA of tobacco etch virus: further characterization and detection of protein linked to RNA. *Virology* **112**: 391-399.

Hari, V., Siegel, A., Rozek, C., and Timberlake, W.E. (1979). The RNA of tobacco etch virus contains poly (A). *Virology* **92:** 568-571.

Hiebert, E., Purcifull, D., and Christie, R.C. (1984). Purification and immunological analysis of plant viral inclusion bodies. Methods in Virology **7**: 225-280., Eds. K. Maramorosh and H. Koprowski. Academic Press, N.Y.

Hellmann, G.M., Shaw, J.G., and Rhoads, R.E. (1988). In vitro analysis of tobacco vein mottling virus NIa cistron; evidence for a virus-encoded protease. *Virology* **163**, 554-562.

Hollings, M., and Brunt, A.A. (1981). Potyviruses. in "Handbook of Plant Virus Infections and Comparative Diagnosis". (Ed. E. Kurstak). pp701-754. Elsevier/North-Holland Biomedical, New York.

Knuhtsen, H., Hiebert, E., and Purcifull, D.E. (1974). Partial purification and some properties of tobacco etch virus intranuclear inclusions. *Virology* **61**: 200-209.

Krausslich, H.-G., and Wimmer, E. (1988). Viral Proteinases. *Annual Rev. Biochem.* **57**, 701-754.

Lain, S., Reichmann, J.L., Mendez, E., and Garcia, J.A. (1988). Nucleotide sequence of the 3' terminal region of plum pox potyvirus RNA. *Virus Research* **10**: 325-342.

Langenberg, W.G. (1986). Virus protein associated with cylindrical inclusions of two viruses that infect wheat. *J. Gen. Virology* **67:** 1161-1168.

Maiss, E., Timpe, U., Brisske, A., Jelkmann, W., Casper, R., Himmler, G., Mattanovich, D., and Katinger, H.W.D. (1989) The complete nucleotide sequence of plum pox virus RNA. *J. Gen. Virology* **70**: 513-524.

Murphy, J., Shaw, J.G., and Rhoads, R.E. (1989). The VPg of tobacco etch virus. Phytopathology 79: 1214 (Abst.)

Nicklin, M.J.H., Toyoda, H., Murray, M.G., and Wimmer, E. (1986). Proteolytic processing in the replication of polio and related viruses. *Bio/Technology* **4,** 33-44.

Oh, C.-S., and Carrington, J.C. (1989). Identification of essential residues in potyvirus proteinase HC-PRO by site-directed mutagenesis. *Virology*. in press.

Ravelonandro, M., Varveri, C., Delbos, R., and Dunez, J. (1988). Nucleotide sequence of the capsid protein gene of plum pox potyvirus. *J. Gen Virology* **69:** 1509-1516.

Robaglia, C., Durand-Tardif, M., Tronchet, M., Boudazin, G., Astier-Manifacier, S. and Casse-Delbart, F. (1989). Nucleotide sequence of potato virus Y (N strain) genomic RNA. *J. Gen. Virology* **70**: 935-947.

Shahabuddin, M., Shaw, J.G., and Rhoads, R.E. (1989). Mapping of the tobacco vein mottling virus VPg cistron. *Virology* **163**: 635-637.

Siaw, M.F.E., Shahabuddin, M., Ballard, S., Shaw, J.G., and Rhoads, R.E. (1985) Identification of a protein covalently linked to the 5'terminus of tobacco vein mottling virus RNA. *Virology* **142**: 134-143.

Slade, D.E., Johnston, R.E., and Dougherty, W.G. (1989). Generation and characterization of monoclonal antibodies reactive with the 49kDa proteinase of tobacco etch virus. *Virology* **173**: in press.

Wellink, J., and van Kammen, A. (1988). Proteases involved in the processing of viral polyproteins. *Arch. Virol.* **98**, 1-26.

Vance, V.B., and Beachy, R.N. (1984a). Translation of soybean mosaic virus RNA in vitro: evidence of protein processing. *Virology* **132**: 271-281.
Vance, V.B., and Beachy, R.N. (1984b). Detection of genomic-length soybean mosaic virus RNA on polyribosomes of infected soybean leaves. *Virology* **138**: 26-36.
Yeh, S.-D, and Gonsalves, D. (1985). Translation of papaya ringspot virus RNA in vitro: detection of a possible polyprotein that is processed for capsid protein, cylindrical-inclusion protein and amorphous-inclusion protein. *Virology* **143**: 260-271.
Ypma-Wong, M.F., Dewalt, P.G., Johnson, V.H., Lamb, J.G., and Semler, B.L. (1988a). Protein 3CD is the major poliovirus proteinase responsible for cleavage of the P1 capsid precursor. *Virology* **166**, 265-270.
Ypma-Wong, M.F., Filman, D.J., Hogle, J.M., and Semler, B.L.(1988b). Structural domains of the poliovirus polyprotein are major determinants for polyprotein cleavage at gln-gly pairs. *J. Biol. Chem.* **263**, 17846-17856.

DISCUSSION OF W. DOUGHERTY'S PRESENTATION

T. Hohn: Two types of auto-processing may occur, i.e., in neighboring molecules of the same type or within a single molecule. Do you have any idea what happens in your system?

W. Dougherty: We made a precursor in which the 49 K proteinase was inactivated by mutagenesis. When we co-translated this precursor with a transcript which synthesized an active proteinase we did not observe any *trans* processing. So it appears that the 49 K proteinase must release itself out of its own polyprotein. Even when you provide lots of protease *in trans* there is no processing. So *in cis* processing is an obligatory reaction.

J. Hammond: I know you have looked at the cleavage sites in a number of different TEV isolates, at least for the NIb-coat protein junction. Do any of the differences in these proteolytic processing sites reflect on the comparative efficiency of reproduction of these strains?

W. Dougherty: No, in fact when we compared different isolates of TEV we only found one case where a single amino acid was changed in one position within the central seven amino acid sequence that defines a cleavage site. In general if you look at any strain the same seven amino acid sequence will be at that particular cleavage site as is found in all the other strains.

J. Hammond: Do you see differences between the reproduction of different strains?

W. Dougherty: We never looked at that in a really careful fashion.

M. Wilson: Have you found any sites in the polyprotein where your seven amino acid target sequence will not be cleaved? You would expect the polyprotein to fold as it is being synthesized. It strikes me that the two places where you dropped the sequence in the witnessed cleavage is marvellous, but have there been sites that you cannot access?

W. Dougherty: When Jim Carrington and I did these experiments we had indeed wondered whether this could be expected, but you should realize that whatever the structure was before, any time that you drop in a seven amino acid sequence you will probably totally disrupt it.

M. Wilson: But you need not necessarily expect it to be on the surface and accessible for the 49 K proteinase.

W. Dougherty: True, but it seems when you now put this cleavage site in some place, or, alternatively, when you move the proteinase sequence around, it still seems to be able to cleave itself out or still seems to recognize the cleavage sites and process it.

M. Wilson: Have you dropped a cleavage site in completely alien genes, not related to TEV?

W. Dougherty: We have hooked on a *Xenopus* sequence (obtained from Mike Bevan), downstream of the 49 K protein sequence, and this foreign sequence is correctly processed out *in vitro*.

R. Goldbach: Juan Garcia and co-workers (see Virus Research **13** (1989) 157-172) have compared the sequences of plum pox virus (PPV), TEV and TVMV and identified a possible sixth proteolytic cleavage site in the polyprotein of all three viruses, upstream of the CI-protein sequence. What is your opinion about this possible cleavage site?

W. Dougherty: The sequence at that position is not really a well-conserved sequence with respect to their matches of other cleavage sites. Since these authors haven't shown actual processing at that site it is still pure computer speculation. For TEV we don't have evidence for any processing at that particular region.

R. Hull: Have you done a computer search of plant genes for cleavage site sequences and is there any evidence that normal plant proteins get processed?

W. Dougherty: We did a computer search to identify all cleavage sites or to identify sites that would have the conserved positions only, that is a glutamic acid at -6, a tyrosine at -3, and a glutamine-serine or glutamine-glycine pair. Screening various data banks we found eighteen such sequences, five of them were from TEV, the other thirteen were found anywhere in proteins, ranging from complement fixing agents to storage proteins of rice and to mammalian enzymes. We screened about 10,000 different proteins, hence it is not a very common sequence.

R. Goldbach: The CPMV proteinase does cleave a rabbit protein in the reticulocyte lysate. Do you observe any such cleavage when translating TEV-RNA *in vitro*?

W. Dougherty: We never noticed this, but we never really looked.

T. Hohn: Are there any site inhibitors?

W. Dougherty: We tested a lot of possible inhibitors, most compounds having none or a marginal effect on the activity. The only compounds that were effective were zinc (5 Mm), iodo-acetamide and other alkylating agents. We have started a project, together with Bud Ryan (Washington State University), to mutate in our cleavage site into a trypsin I inhibitor.

R. Goldbach: The 3C-proteinases of picornaviruses and the related proteinases of como- and potyviruses are in fact all serine-type proteinases in which the serine of the active site has been replaced by a cysteine residue. Do you know of any cellular serine-type proteinase which also has a replacement to cysteine and do you know what this replacement in the viral proteinases mean?

W. Dougherty: No.

SUMMARY AND CONCLUDING REMARKS BY THE THIRD SESSION CHAIRMAN, ROB GOLDBACH

R. Goldbach: The contributions of John Shaw and Bill Dougherty both demonstrate that during the past 10 years we have gained considerable insight into the molecular and pathological properties of the most important group of plant viruses, the potyviruses. Hence, complete nucleotide sequences of the genomes of four different members, i.e., TEV, TVMV, PVY and PPV, have been published to date. In addition, since in view of their great economical impact a considerable number of virologists is currently applying the now well established method of "coat protein mediated resistance" (see the contribution of R.N. Beachy and co-workers in this volume), partial sequence data of maybe a few dozen other potyviruses have become available, in particular coat protein gene sequences. Furthermore, we have gained increased knowledge about the functions of the various proteins encoded by the potyviral genome, as discussed in both contributions.

R. Goldbach: The presence of a proteinase domain in the helper component (as a consequence now often referred to as HC-PRO) of TEV, as reported by Dougherty and co-workers, is really an important and intriguing finding. The question now raised is whether this proteolytic activity is involved not only in release of the HC from the polyprotein (at its C-terminal side) but also in the functioning of this protein during virus transmission.

R. Hull: The HC obviously enables the aphid to acquire the virus but for transmission the particle should get released again. Possibly, under the conditions within the aphid's stylet the protease domain operates on the HC and cuts it to release the virus.

B. Harrison: Alternatively, release may occur by a proteolytic cleavage event in the coat protein sequence. Robinson and I have pointed out (Harrison, B.D. and Robinson, D.J., Proceedings of the Royal Society of London (1988) Series B, Volume **321**, p. 447-462) that the N-terminal part of the coat protein, which is very variable among different potyviruses, usually contains a triplet of amino acids, DAG. Furthermore, downstream of this conserved sequence there is a highly conserved trypsin cleavage site (usually KDKD or KDRD) and the idea we put forward was that the DAG could be the site where the virus binds to the HC and that release occurs by cleavage by a trypsin-like enzyme, which leaves the N-terminal part bound and releases the virus particle.

T. Pirone: Certainly in the aphid's food canal there are all sorts of enzymes and such an enzyme would indeed not be very unusual. Furthermore, we have just finished sequencing the potato virus C isolate of PVY, which produces a HC that is not biologically active. Hence, we may be able to study some of these questions by making mutations in areas which might have biological relevence to the activity of the HC.

J. Carrington: I would like to make the proposition that HC-PRO would be not involved in releasing the virus particle by proteolytic cleavage at the trypsin-like cleavage site.

R. Goldbach: To continue with the cylindric inclusion (CI) protein, this protein, as mentioned by John Shaw, contains an NTP-binding motif and might have helicase activity, but this still needs to be demonstrated. According to Gorbalenya *et al.*, [FEBS Letters **235**, 16-24 (1988)] the rationale for a plus strand RNA virus to encode a helicase could be that such activity is required for unwinding the RF molecules during replication, to enable the RNA polymerase to produce progeny plus strand molecules. For PPV the NIa protein has been shown to have ATPase-activity (Lain *et al.*, Homologous poty-, flavi- and pestivirus proteins belonging to a superfamily of helicase-like proteins. RNA stimulated ATPase activity of plum pox potyvirus CI protein. (Abstract 2nd International Symposium on Positive Strand RNA Viruses, June 26-30, 1989, Vienna, Austria), which is not in disagreement with an unwinding activity.

Concerning the NIa protein, i.e., the viral proteinase, Bill Dougherty and co-workers have presented detailed and very worthy information with respect to cleavage site requirements, and the active site and putative substrate binding pocket of this protein. The progress in our knowledge on this viral function has been marvellous and the potyviral proteinase ranks to date among the best characterized viral proteinases.

As discussed by John Shaw, the NIb protein must represent the virus RNA-dependent RNA polymerase. This protein exhibits sequence homology to other viral polymerases but direct biochemical evidence for having RNA synthesing activity has not been published yet.

W. Dougherty: We have been able to prepare extracts that can incorporate nucleotides into TCA-precipitable products and this activity can be specifically inhibited with antiserum to NIb, but not with antiserum to CI or the coat protein. Furthermore, we have made a number of constructs, that are not tested yet, in which we have cloned the polymerase gene. We will try to express and study this protein in a prokaryotic system, similar to what has been done for the polioviral enzyme.

R. Goldbach: Both the NIa and NIb proteins as well as the CI protein are often accumulated as characteristic inclusions. The question as to whether these inclusions just represent proteins which have become inactive after one round of replication and are then deposited as junk in for example the cell nucleus, or whether these inclusions carry out certain specific functions during the viral multiplication cycle still remains to be answered.

In addition to the proteins discussed above, potyviruses specify a number of non-structural proteins of which any information about their function is completely lacking. Hence, for fully understanding the expression and replication strategy of potyviruses, there is still a long way to go.

As far as the potyviral coat protein, Shukla and co-workers especially (see for instance Shukla and Ward, Archives of Virology **106** (1989) 171-200) have done detailed sequence comparisons which provide very useful data with respect to the taxonomy of potyviruses. The potyviral coat proteins appear to be characterized by having a very conserved core protein sequence (with approximately 40-70% sequence homology among all potyviruses compared thus far) and a unique N-terminus, which may contain important determinants, for example, for host range and aphid transmission.

Other important achievements in potyvirus research during the past few years, and quoted by Shaw, have been the development and availability of a number of important tools that will effectively enable study of these viruses in even greater detail. These are (i) a tobacco protoplast system in which by electroporation a potyvirus infection can be established and consequently studied (Luciano *et al.*, Plant Science **51** (1987) 295-303) and (ii) a full-length cDNA clone of TVMV RNA from which infectious transcripts can be derived (Domier *et al.*, PNAS **86** (1989) 3509-3513). This full-length clone will allow both site-directed mutagenesis of the potyviral genome and complementation studies with transgenic plants expressing viral proteins.

Last but not least John Shaw presented the first results obtained with transgenic plants expressing potyviral proteins, which will permit study of the role of these proteins, e.g., the inclusion body proteins, in processes like symptom induction. Additionally, a growing number of virologists are currently trying to obtain coat protein mediated resistance for potyviruses (which is not unexpected, in view of their economical impact). In general it seems, however, that for potyviruses the levels of coat protein expressed in transgenics are relatively low compared to other viruses. Possibly this is due to a higher instability of the potyviral coat protein in such plants. Indeed the poster presented by Kaniewski and co-workers at this meeting (Kaniewski *et al.*, Field performance of PVX and PVY resistant potatoes), showing results with potatoes expressing the coat proteins of PVX (strongly) and PVY (weakly) seems to support this idea.

N. Tumer: We don't know if the low level of expression of PVY coat protein in transgenic potato is due to instability of this protein or to low translation efficiency. In Roger Beachy's group the expression of SMV coat protein in transgenic tobacco was measured to be at levels similar to those reported for TMV coat protein. In our case we worked specifically with PVY coat protein and that has been expressed at much lower levels in general.

J. Carrington: Could this be due to the 5'-end of the mRNA in the case of the PVY construct?

N. Tumer: The 5'- end had about 11 nucleotides from the authentic PVY sequence upstream of the coat protein cistron and this sequence was fused to the CaMV 35S leader sequence.

J. Carrington: The level of potyviral proteins expressed in transgenic plants can be increased by fusing the 5'-nontranslated region of the potyviral RNA to the cistron to be expressed. Using reporter constructs the level of expression may increase 5 to 200 fold depending on the plant species.

M. Wilson: Alternatively, the TMV leader sequence could be very useful. In experiments where several groups try to express a gene which doesn't have a good leader sequence itself the TMV leader sequence appears to enhance translations in a very positive way. In particular the TMV omega leader sequence could be useful. This sequence, cloned in a plasmid and obtainable from me on request, can be used with benefit in this regard.

Molecular Biology of Bromovirus Replication and Host Specificity

Paul Ahlquist, Richard Allison, Walter Dejong, Michael Janda, Philip Kroner, Radiya Pacha and Patricia Traynor

Institute for Molecular Virology and Department of Plant Pathology
University of Wisconsin - Madison, Madison, WI 53706, USA

One of the remarkable developments in virology in recent years has been the recognition that many outwardly diverse RNA and DNA viruses of plants, vertebrate animals, insects, and even bacteria share extensive similarities in nonstructural proteins involved in nucleic acid replication. The emergence of these revelations from nucleic acid sequencing has enlightened and stimulated considerations of virus evolution and has accelerated research progress by allowing productive comparison of observations from distinct but related viruses. Among (+) strand RNA viruses, which comprise the majority of known plant viruses, such relationships have defined at least two and possibly three virus superfamilies (Goldbach, 1987). One of these superfamilies contains many of the well studied viruses encapsidating capped (+) strand RNAs (Cornelissen and Bol, 1984; Haseloff *et al.*, 1984; Ahlquist *et al.*, 1985). As examples, Fig. 1 shows the relationship among genome maps of three viruses in this superfamily: brome mosaic virus (BMV), tobacco mosaic virus (TMV), and Sindbis virus, type member of the animal alphaviruses. As shown, separate domains in nonstructural proteins nsP1, nsP2 and nsP4 of Sindbis virus share extensive amino acid conservation with nonstructural proteins 1a and 2a of BMV and the 126/183 kD proteins of TMV. By a variety of studies all of these proteins have now been implicated in RNA replication. Thus in all three viruses most of the genome encodes RNA replication functions, while additional genes for nonstructural proteins and capsid proteins complete the ability of each virus to move from cell to cell and host to host. Many other plant and animal viruses encode proteins similar to one, two, or all three of the conserved domains mapped in Fig. 1.

From this perspective, many important questions dealing with these viruses can be meaningfully grouped under two headings: First, what functions do the conserved nonstructural proteins encode, and what is the complete mechanism of the genome replication cycle driven by this highly conserved set of RNA replication genes? Second, what differential characteristics are responsible for the specialized adaptation of each virus to a particular set of hosts? To address these issues, we have studied aspects of RNA replication, gene expression, and host specificity in BMV, the type member of the bromoviruses, and cowpea chlorotic mottle virus (CCMV), its closest known relative. BMV and CCMV are adapted to systemically infect cereals and legumes, respectively, and in this and other respects present an interesting and useful mixture of similarities and differences.

ORGANIZATION AND GENETIC MANIPULATION OF BROMOVIRUS GENOMES

BMV and CCMV share many general properties of genome and virion structure (Lane, 1981). The genome of each virus is divided among three RNAs, designated RNA1 (3.2 kb), RNA2 (2.9 kb), and RNA3 (2.1 kb) (Fig. 1). Monocistronic RNAs 1 and 2 encode proteins 1a (109 kD) and 2a (94 kD), which are *trans*-acting RNA replication factors. RNA3 encodes the nonstructural 3a protein (32 kD) and the coat protein (20 kD). Coat protein is expressed by the production of a subgenomic mRNA, RNA4. We have sequenced the entire BMV genome (Ahlquist *et al.*, 1981b; 1984b) and CCMV RNAs 2 and 3 (Allison *et al.*, 1989). CCMV RNA1 has been sequenced by Jozef Bujarski (personal communication). The BMV and CCMV sequences are actually somewhat more divergent than had been expected. When compared between the two viruses, the 1a, 2a, 3a and coat proteins differ at approximately 20, 40, 50, and 30% of their amino acid positions, respectively, after allowing for various small gaps in the aligned sequences. In addition, there are some significant sequence variations and insertion/deletion differences in the noncoding regions. Most notable among these is that, just 5' to the 3a gene, CCMV RNA3 contains an 111 base insertion with respect to the corresponding region of BMV RNA3 (Allison *et al.*, 1989). The 3' 200 bases of RNAs 1-3 are highly conserved within each virus and form an extensive secondary structure that is similar in BMV and CCMV (Ahlquist *et al.*, 1981a). These 3' ends carry out a number of tRNA-specific reactions including aminoacylation and encode signals for (-) strand RNA synthesis (Bujarski *et al.*, 1986; Dreher *et al.*, 1988).

Both BMV and CCMV can be readily manipulated at the level of cloned cDNA. BMV was the first RNA virus for which infectious *in vitro* transcripts were produced from cloned cDNA (Ahlquist *et al.*, 1984c). More recently, we have also produced infectious *in vitro* transcripts from CCMV cDNA clones (Allison *et al.*, 1988). In addition to facilitating studies of the individual viruses, recombinant DNA manipulation of such biologically active cDNA clones can be used to make experimentally useful sequence exchanges between viruses. The feasibility of such an approach was previously demonstrated by the insertion of nonviral genes in BMV (French *et al.*, 1986) and by the construction of hybrids between relatively disparate viruses such as BMV and TMV (Sacher *et al.*, 1988). Recent experiments show that an even larger array of functional virus hybrids can be made by transferring selected genes, gene segments and regulatory sequences between BMV and CCMV. Such exchanges have the potential to be extremely informative since current experience indicates that BMV and CCMV are both sufficiently related for successful

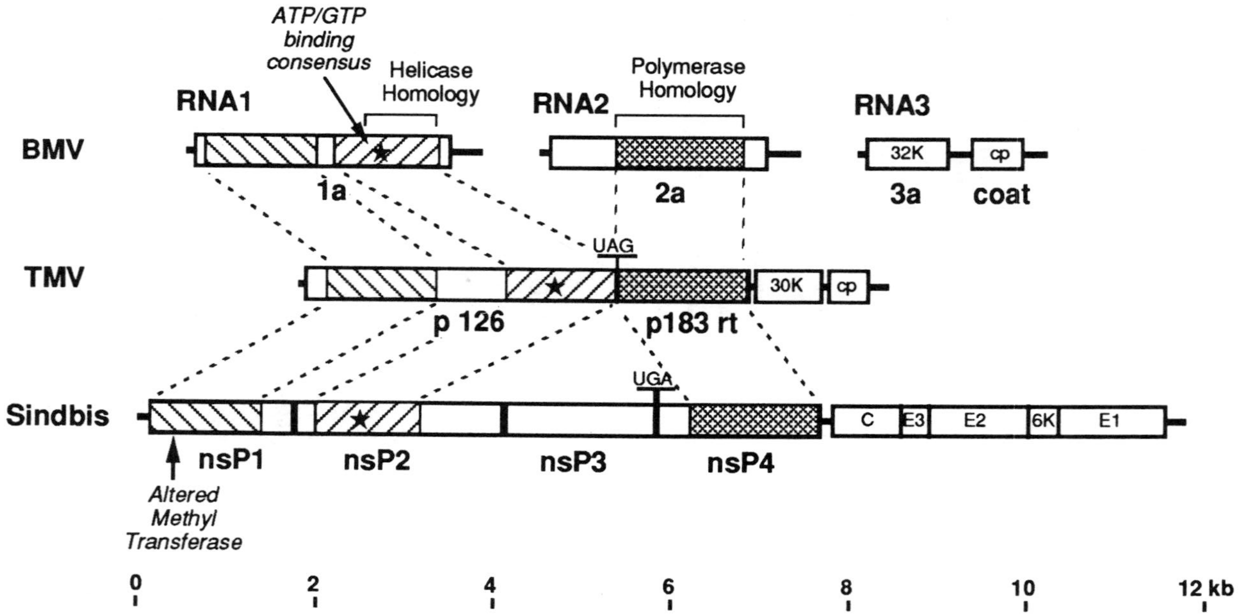

Fig.1 Schematic diagram of the (+) strand RNA genomes of BMV, TMV and Sindbis virus, an animal alphavirus. Boxed regions indicate open reading frames, and the three shaded domains within these open reading frames denote the extent of amino acid sequences conserved among the encoded proteins (Haseloff *et al.*, 1984; Ahlquist *et al.*, 1985). Readthrough termination codons in TMV and Sindbis virus are marked, and p183 rt denotes the readthrough portion of the TMV 183 kD protein. See text for further details.

genetic exchange and sufficiently diverged to possess altered specificities in a number of crucial molecular interactions of viral replication and host interaction.

FUNCTIONS AND POSSIBLE INTERACTIONS OF BROMOVIRUS 1A AND 2A PROTEINS

BMV RNAs 1 and 2 replicate in protoplasts in the absence of RNA3 (Kiberstis *et al.*, 1981), and the 3a and coat genes encoded on RNA3 are further dispensable for replication of RNA3 itself and for synthesis of subgenomic RNA4 (French and Ahlquist, 1987). Both RNAs 1 and 2 are required to direct RNA replication (French *et al.*, 1986), and frameshift or other mutations in the 1a (P. Kroner, unpublished results) or 2a genes (Traynor and Ahlquist, 1990) block viral RNA synthesis. The 1a and 2a genes are thus the only virus-encoded factors required *in trans* for RNA replication.

Possible functions for the bromovirus 1a and 2a proteins and their homologues in other viruses have been suggested by further sequence comparisons. The conserved central domain of the 2a protein has extensive similarity to poliovirus RNA-dependent RNA polymerase 3D and the ß-subunit of Qß RNA replicase (Kamer and Argos, 1984) and lesser similarities to the RNA polymerase of dsRNA reoviruses and birnaviruses (Gorbalenya and Koonin, 1988; Morosov, 1989) and to reverse transcriptase (Argos, 1988). Similar polymerase-like domains are encoded by the animal flaviviruses, coronaviruses, and nodaviruses and many RNA plant viruses (Goldbach, 1987). The C-terminal conserved domain of protein 1a contains a purine nucleotide binding consensus found in ATPases and GTPases and extended similarities to a number of ATP-dependent bacterial helicases and putative herpesvirus helicases (Hodgman, 1988). While 1a-related domains encoded by the animal alphaviruses and coronaviruses and many plant viruses share this relation, proteins of the animal flaviviruses and pestiviruses and the plant potyviruses have similarities to a related but distinct group of helicases (Gorbalenya *et al.*, 1989). Finally, the Sindbis nsP1 domain, which is related to the N-terminal portion of 1a, is the site of a mutant affecting a methyltransferase activity implicated in viral RNA capping (Mi *et al.*, 1989).

To more directly explore the role of bromovirus proteins 1a and 2a in RNA replication, we have introduced a number of substitution and linker insertion mutants throughout both genes in BMV (Kroner *et al.*, 1989; P. Kroner, P. Traynor and P. Ahlquist, unpublished results). These mutations have produced a wide variety of RNA replication phenotypes, and the distribution of these effects within the genes confirms that all three conserved domains of Fig. 1 are involved in RNA replication. Many mutants in the central domain of protein 2a, e.g., produce complete, partial, or temperature sensitive blocks to synthesis of all viral RNA species, consistent with the suggested role of 2a in viral RNA elongation (Kroner *et al.*, 1989). Some mutants in this conserved 2a domain, however, alter the ration of genomic to subgenomic RNA, suggesting that the 2a protein could also be involved in regulating the initiation of specific classes of viral RNA (ibid.).

Although BMV and CCMV systemically infect different hosts, both viruses replicate in protoplasts from barley (Allison *et al.*, 1988) as well as from a number of other host and nonhost plants. Interestingly, heterologous combinations of BMV and CCMV RNAs 1 and 2 fail to direct detectable RNA replication, suggesting that the heterologous combinations of the encoded 1a and 2a proteins may be functionally incompatible (Allison *et al.*, 1988). This is a potentially important issue because such incompatibility would indicate that successful RNA replication requires direct or indirect interaction of the 1a and 2a proteins. To localize the determinants of compatible and incompatible interaction between BMV and CCMV RNAs 1 and 2, a number of hybrids were made by exchanging selected segments between BMV and CCMV RNA2 (Traynor and Ahlquist, 1990). Despite the substantial divergence of the BMV and CCMV 2a proteins, many of the hybrid 2a proteins were functional and supported RNA replication in combination with RNA1 from BMV or CCMV. Unlike either wild type 2a gene, some hybrid 2a proteins supported RNA replication *in trans* with either BMV or CCMV RNA1. The overall results of these exchanges indicated that successful interaction with BMV or CCMV RNA1 was largely controlled by N- and C-terminal segments of the 2a gene.

These results are consistent with the possibility that, for at least some steps in RNA replication, the 1a and 2a proteins may be part of a single replication complex. Such a 1a/2a complex would be similar to the TMV 183 kD readthrough protein, which covalently joins 1a- and 2a-like domains in a single protein (Fig. 1). A related complex might form from some or all of the Sindbis virus nonstructural proteins, which are proteolytically processed from a single polyprotein. In keeping with the hybrid 2a gene results, association of the bromovirus 1a and 2a proteins might be mediated by the nonconserved extensions flanking the polymerase-like 2a protein core. Corresponding extensions are absent in TMV, where the polymerase-like domain is fused directly to the C-terminus of the 1a-like 126 kD protein (Fig. 1). Association of the 2a polymerase-like domain with the helicase-like 1a domain would account for the observed ability of an RNA-dependent RNA polymerase extract from BMV-infected cells to copy partially double stranded BMV templates *in vitro* (Ahlquist *et al.*, 1984a). In view of these possibilities, further experiments are in progress to assess the relative contribution of protein-protein and protein-RNA interactions to the compatibility or incompatibility of various RNA1 and RNA2 combinations.

TEMPLATE SPECIFICITY AND CIS-ACTING RNA REPLICATION SIGNALS

Viral RNA replication is not only dependent on *trans*-acting factors such as the 1a and 2a proteins, but also on *cis*-acting replication signals that direct these factors to their appropriate template RNAs. *In vivo*, BMV and CCMV RNA replication show differences in template specificity which can be used to study such signals and the determinants of their recognition. While BMV RNA3 is replicated to a high level in cells co-inoculated with CCMV RNAs 1 and 2, e.g., CCMV RNA3 is only poorly amplified in cells co-inoculated with BMV RNAs 1 and 2 (Allison *et al.*, 1988). In studies with the RNA2 hybrids discussed above, none of the segments exchanged between BMV and CCMV RNA2 affected RNA3 recognition. However, for each of two RNA2 hybrids able to function in combination with either BMV or CCMV RNA1, this BMV- and CCMV-specific difference in RNA3 template recognition segregated with RNA1 (Traynor and Ahlquist, 1990). At least some aspects of template selection thus appear to be controlled by the 1a protein.

On the converse side of template specificity, *cis*-acting replication signals have been mapped by deletion analysis in both BMV and CCMV RNA3. Efficient amplification of BMV RNA3 requires not only 5'- and 3'-terminal sequences, but also a 95-150 base element from the intercistronic noncoding region (French and Ahlquist, 1987). Though located over 1 kb from either end of RNA3, deletions into this region inhibited RNA3 amplification 100-fold. Since 3'-terminal sequences are sufficient to direct (-) strand synthesis *in vitro*, this region may contribute to (+) strand synthesis or viral RNA stability. This required intercistronic region contains the conserved sequence GGUUCAAyCCCU, which is also present near the 5' ends of BMV, CCMV and cucumber mosaic virus RNAs (Rezaian *et al.*, 1985; French *et al.*, 1987; Allison *et al.*, 1989). This segment corresponds to a consensus element in cellular RNA *pol III* promoters (Marsh and Hall, 1987) and to the invariant residues of the TΨC loop of tRNAs. Together with the tRNA-like 3' ends of bromovirus RNAs, this suggests that cellular as well as viral factors might interact with viral RNA to facilitate or modulate RNA synthesis.

We have similarly defined replication signals in CCMV RNA3 (Pacha *et al.*, 1989). Unlike BMV RNA3, the CCMV RNA3 intercistronic region is dispensable for replication, showing that RNA replication signals may be organized in different ways in even closely related viruses. Such genetic flexibility may underlie much of the evolutionary divergence in genome organization seen among viruses with replication genes related to BMV (Fig. 1). Interestingly, the dispensable CCMV RNA3 intercistronic segment lacks the *pol III* promoter consensus element found in the corresponding BMV RNA3 region (Allison *et al.*, 1989).

Another set of *cis*-acting elements in viral RNA synthesis which have been studied both *in vitro* (Marsh *et al.*, 1988) and *in vivo* (French and Ahlquist, 1988) are the sequences which direct production of BMV subgenomic RNA4. RNA4 is produced by internal initiation of RNA synthesis on (-) strand RNA3 (Miller *et al.*, 1985). *In vivo* studies reveal that the promoter sequences which direct this initiation are surprisingly extended and complex (French and Ahlquist, 1988). Low levels of correctly initiated subgenomic RNA synthesis are directed by a core promoter segment extending only 20 bases upstream and

16 bases downstream of the RNA4 start site. However, sequences extending 74 to 95 bases upstream of the start site are required for wild type levels of RNA4 synthesis. This region upstream of the "core promoter" includes three elements whose individual loss inhibits subgenomic RNA production: an 18 base oligo (A) tract, an imperfect upstream repeat of the core promoter, and a partial upstream repeat of the core promoter. Most interestingly, engineered duplications creating additional upstream copies of the repeat elements stimulate promoter activity above wild type levels.

Exchange of RNA3 between BMV and CCMV shows that the RNA synthesis machinery of each virus recognizes the subgenomic promoter of the heterologous RNA3 (Allison *et al.*, 1988). In keeping with this, the region immediately upstream of the CCMV coat gene is highly similar to the BMV core promoter sequence and is also preceeded by an oligo(A) tract. Unexpectedly, the 3' end of the 111-base "insertion" preceeding the CCMV 3a gene (see above) also shows significant similarity with the core subgenomic promoter. The positioning of a subgenomic promoter-like sequence 5' to this gene has possible evolutionary significance since it is consistent with derivation of CCMV RNA3 or its 3a gene from an ancestral virus with fewer genomic RNAs (Allison *et al.*, 1989).

BROMOVIRUS RNA ENCAPSIDATION

Bromovirus virions are nonenveloped particles whose approximately 28 nm diameter icosahedral capsid is composed of 180 coat protein subunits. The three genomic RNAs are separately encapsidated, giving rise to three virion classes with different densities (Lane, 1981). Particles in the heaviest and lightest classes contain one copy of RNA1 or RNA2, respectively, while middle density particles each contain one copy of RNA3 and one copy of subgenomic RNA4. Protoplast experiments show that BMV and CCMV coat proteins readily encapsidate the RNAs of the heterologous virus (Allison *et al.*, 1988). This mutually permissive encapsidation simplifies the construction of hybrids between the two viruses (see below).

Bromovirus coat proteins, like those of many other isometric RNA viruses, have highly basic N-terminal segments. A variety of *in vitro* studies with both BMV and CCMV suggest that these basic N-terminal segments interact with viral RNA in the final particle (Sgro *et al.*, 1986; Vriend *et al.*, 1986). Recent experiments engineering designed changes into this region of the BMV coat protein gene confirm its importance for RNA packaging *in vivo* and reveal other coat protein effects (Sacher and Ahlquist, 1989): Removal of the N-terminal 25 aa, which contains all eight N-proximal basic residues, gives near normal coat protein accumulation *in vivo* but completely blocks RNA packaging. By contrast, the first seven aa are completely dispensable for packaging and full systemic infection even though these residues include the first charged amino acid and are significantly conserved between BMV and

CCMV. A mutation reducing coat protein production 10-fold blocks all detectable packaging, suggesting that coat protein must accumulate to a certain threshold for effective encapsidation. Finally, the production of nonstructural proteins is stimulated in infections with mutants failing to express encapsidation-competent coat protein, suggesting that encapsidation may provide a simple level of gene regulation in bromovirus infection.

HOST SPECIFICITY

As with many plant viruses, little evidence of host specificity is seen at the level of intracellular replication events in bromovirus infection: CCMV directs RNA synthesis and packaging in protoplasts from the nonhost plant, barley, while BMV replicates in cells from a wide variety of host and nonhost plants. Practical differences in BMV and CCMV host specificity appear to be determined largely by the competence of each virus to systemically infect particular host plants. Currently the processes of virus movement and systemic infection in plants are very poorly understood, as are host resistance and defense mechanisms. Beyond RNA replication and packaging, requirements for successful systemic infection may include virus functions or adaptations to support infection spread beyond primarily infected cells and/or to avoid or suppress host defenses. Host specificity may thus derive from differences in the success of virus-host interactions at these and other presently unrecognized levels.

The genetic distance between the monocotyledonous systemic hosts of BMV and the dicotyledonous systemic hosts of CCMV makes it likely that these two viruses necessarily differ in many of the virus-host interactions that affect successful systemic infection. As in replication studies, this degree of divergence makes the system more challenging in some respects, but also suggests that combined BMV/CCMV studies should support exploration of multiple facets of virus-host interaction.

It is already clear that adaptation of BMV and CCMV to barley and cowpea, respectively, is not controlled by a single gene but rather by at least two and possibly three genes. This is demonstrated, e.g., by the behaviour of the reassortant genomes made by exchanging RNA3 between BMV and CCMV (Allison *et al.*, 1988). Both of these reassortants replicate and encapsidate all of their RNAs in protoplasts and give local lesion infections on *Chenopodium hybridum*, a local lesion host for both BMV and CCMV. However, neither of these reassortants infects either barley or cowpea, the normal hosts of the parent viruses. The loss of adaptation to normal systemic hosts upon transfer of the heterologous RNA3 into each genome indicates that appropriate adaptation of some factor or factors encoded by RNA3 must be required for successful systemic infection of barley and cowpea. However, systemic infection must also require proper adaptation of factors encoded by RNA1 and/or RNA2; if this were not true, transferring RNA3 between the two viruses would transfer host range.

The presence of host specificity determinant(s) on bromovirus RNA3 is not unexpected since the functions encoded by this RNA are only required at the level of systemic infection. However, the involvement of RNAs 1 and 2 in host specificity does not obviously follow from other known bromovirus characteristics. As noted above, the only established functions encoded by RNAs 1 and 2 are those of RNA replication, a process that appears to transcend the host specificity differences of BMV and CCMV. These considerations imply that RNAs 1 and 2 may encode some additional functions other than those involved in RNA replication, or that host specificity may involve some unrecognized, possibly subtle characteristics of RNA replication that have not been revealed in protoplast replication studies performed to date.

Although some plant RNA viruses are able to spread systemically without encapsidation (Hamilton and Baulcombe, 1989; Dawson *et al.*, 1988), recent experiments indicate that expression of encapsidation-competent coat protein is required for systemic infection by BMV (Sacher and Ahlquist, 1989) and CCMV (R. Allison, unpublished results). Other results show that the 3a gene is required for systemic infection (R. Allison, R. Sacher and P. Ahlquist, unpublished results). Thus, all four bromovirus genes are required for systemic infection, and the contribution of each to host specificity is currently being characterized. In one approach, BMV/CCMV hybrids functional at a variety of levels in infection have been constructed and are being tested in several different hosts (R. Allison, M. Janda and R. Pacha, unpublished results). Such studies are already revealing significant aspects of the function and adaptation of specific virus genes and are providing important contributions toward defining the virus-host interactions and mechanisms that determine the success or failure of virus infections.

ACKNOWLEDGEMENTS

We thank Ben Young and Craig Thompson for excellent technical assistance in a variety of experiments. This research was supported by the National Institutes of Health under Publich Health Service Grant GM35072 and by the National Science Foundation under Presidential Young Investigator Award DMB-8451884.

REFERENCES

Ahlquist, P., Bujarski, J., Kaesberg, P., and Hall, T. C. (1984a). Localization of the replicase recognition site within brome mosaic virus RNA by hybrid-arrested RNA synthesis. *Plant Mol. Biol.* **3**, 37-44.

Ahlquist, P., Dasgupta, R., and Kaesberg, P. (1981a). Near identity of 3' RNA secondary structure in bromoviruses and cucumber mosaic virus. *Cell* **23**, 183-189.

Ahlquist, P., Dasgupta, R., and Kaesberg, P. (1984b). Nucleotide sequence of the brome mosaic virus genome and its implications for viral replication. *J. Mol. Biol.* **172**, 369-383.

Ahlquist, P., French, R., Janda, M., and Loesch-Fries, L. S. (1984c). Multicomponent RNA plant virus infection derived from cloned viral cDNA. *Proc. Natl. Acad. Sci. USA* **81**, 7066-7070.

Ahlquist, P., Luckow, V., and Kaesberg, P. (1981b). Complete nucleotide sequence of brome mosaic virus RNA3. *J. Mol. Biol.* **153**, 23-38.

Ahlquist, P., Strauss, E., Rice, C., Strauss, J., Haseloff, J., and Zimmern, D. (1985). Sindbis virus proteins nsP1 and nsP2 contain homology to nonstructural proteins from several RNA plant viruses. *J. Virol.* **53**, 536-542.

Allison, R. F., Janda, M., and Ahlquist, P. (1988). Infectious *in vitro* transcripts from cowpea chlorotic mottle virus cDNA clones and exchange of individual RNA components with brome mosaic virus. *J. Virol* **62**, 3581-3588.

Allison, R. F., Janda, M., and Ahlquist, P. (1989). Sequence of cowpea chlorotic mottle virus RNAs 2 and 3 and evidence of a recombination event during bromovirus evolution. *Virology* **172**, 321-330.

Argos, P. (1988). A sequence motif in many polymerases. *Nucleic Acids Res.* **16**, 9909-9916.

Bujarski, J., Ahlquist, P., Hall, T., Dreher, T., and Kaesberg, P. (1986). Modulation of replication, aminoacylation and adenylation *in vitro* and infectivity in vivo of BMV RNAs containing deletions within the multifunctional 3' end. *EMBO J.* **5**, 1769-1774.

Cornelissen, B. J. C., and Bol, J. F. (1984). Homology between the proteins encoded by tobacco mosaic virus and two tricornaviruses. *Plant Mol. Biol.* **3**, 379-384.

Dawson, W., Bubrick, P. and Grantham, G. (1988). Modifications of the tobacco mosaic virus coat protein affecting replication, movement and symptomology. *Phytopathology* **78**, 783-789.

Dreher, T. W., and Hall, T. C. (1988). Mutational analysis of the sequence and structural requirements in brome mosaic virus RNA for minus strand promoter activity. *J. Mol. Biol.* **201**, 31-40.

French, R., and Ahlquist, P. (1987). Intercistronic as well as terminal sequences are required for efficient amplification of brome mosaic virus RNA3. *J. Virol.* **61**, 1457-1465.

French, R. and Ahlquist, P. (1988). Characterization and engineering of sequences controlling *in vitro* synthesis of brome mosaic virus subgenomic RNA. *J. Virol.* **62**, 2411-2420.

French, R., Janda, M., and Ahlquist, P. (1986). Bacterial gene inserted in a engineered RNA virus: Efficient expression in monocotyledonous plant cells. *Science* **231**, 1294-1297.

Goldbach, R. (1987). Genome similarities between plant and animal RNA viruses. *Microbiol. Science* **4**, 197-202.

Gorbalenya, A., and Koonin, E. (1988). Birnavirus RNA polymerase is related to polymerases of positive strand RNA viruses. *Nucleic Acids Res.* **16**, 7735.

Gorbalenya, A., Koonin, E., Donchenko, A. and Blinov, V. (1989). Two related superfamilies of putative helicawses involved in replication, recombination, repair and expression of DNA and RNA genomes. *Nucleic Acids Res.* **17**, 4713-4730.

Hamilton, W. and Baulcombe, D. (1989). Infectious RNA produced by *in vitro* transcription of a full-length tobacco rattle virus RNA-1 cDNA. *J. Gen. Virol.* **70**, 963-968.

Haseloff, J., Goelet, P., Zimmern, D., Ahlquist, P., Dasgupta, R., and Kaesberg, P. (1984). Striking similarities in amino acid sequence among nonstructural proteins encoded by RNA viruses that have dissimilar genomic organization. *Proc. Natl. Acad. Sci. USA.* **81**, 4358-4362.

Hodgman, T. (1988). A new superfamily of replicative proteins. *Nature* **333**, 22-23 and **333**, 578.

Kroner, P., Richards, D., Traynor, P. and Ahlquist, P. (1989). Defined mutations in a small region of the brome mosaic virus 2a gene cause diverse temperature-sensitive RNA replication phenotypes. *J. Virol.* **63** (in press).

Kamer, G., and Argos, P. (1984). Primary structural comparison of RNA-dependent polymerases from plant, animal, and bacterial viruses. *Nucleic Acids Res.* **12**, 7269-7282.

Kiberstis, P., Loesch-Fries, L. S., and Hall, T. (1981). Viral protein synthesis in barley protoplasts inoculated with native and fractionated brome mosaic virus RNA. *Virology* **112**, 804-808.

Lane, L. (1981). Bromoviruses, In *Handbook of Plant Virus Infections and Comparative Diagnosis* (E. Kurstak, Ed.), pp. 333-376. Elsevier Biomedical Press, Amsterdam.

Marsh, L. E., Dreher, T. W., and Hall, T. C. (1988). Mutational analysis of the core and modulator sequences of the BMV RNA3 subgenomic promoter. *Nucleic Acids Res.* **16**, 981-995.

Marsh, L. E., and Hall, T. C. (1987). Evidence implicating a tRNA heritage for the promoters of positive-strand RNA synthesis in brome mosaic and related viruses. *Cold Spring Harbor Symp. Quant. Biol.* **52**, 331-341.

Mi, S., Durbin, R., Huang, H., Rice, C. and Stollar, V. (1989). Association of the Sindbis virus methyltransferase activity with the nonstructural protein nsP1. *Virology* **170**, 385-391.

Miller, W. A., Dreher, T., and Hall, T. (1985). Synthesis of brome mosaic virus subgenomic RNA *in vitro* by internal initiation on (-)-sense genomic RNA. *Nature* **313**, 68-70.

Morozov, S. (1989). A possible relation of reovirus putative RNA polymerase to polymerases of positive-strand RNA viruses. *Nucl. Acids Res.* **17**, 5394.

Pacha, R., Allison, R. and Ahlquist, P. (1989). Cis-acting sequences required for in vivo amplification of genomic RNA3 are organized differently in related bromoviruses. *Virology* (in press).

Rezaian, M. A., Williams, R. H. V., and Symons, R. H. (1985). Nucleotide sequence of cucumber mosaic virus RNA 1. *Eur. J. Biochem.* **150**, 331-339.

Sacher, R. and Ahlquist, P. (1989). Effects of deletions in the N-terminal basic arm of brome mosaic virus coat protein on RNA packaging and systemic infection. *J. Virol.* **63** (in press).

Sacher, R., French, R. and Ahlquist, P. (1988). Hybrid brome mosaic virus RNAs express and are packaged in tobacco mosaic virus coat protein in vivo. *Virology* **167**, 15-24.

Sgro, J., Jacrot, B. and Chroboczek, J. (1986). Identification of regions of brome mosaic virus coat protein chemically cross-linked in situ to viral RNA. *Eur. J. Biochem.* **154**, 69-76.

Traynor, P. and Ahlquist, P. (1990). Use of bromovirus RNA2 hybrids to map cis- and tran-acting functions in a conserved RNA replication gene. *J. Virol.* **64** (in press).

Vriend, G., Verduin, B.J.M. and Hemminga, M.A. (1986). Role of the N-terminal part of the coat protein in the assembly of cowpea chlorotic mottle virus. A 500 MHz proton nuclear magnetic resonance study and structural calculations. *J. Mol. Biol.* **191**, 453-460.

DISCUSSION OF P. AHLQUIST'S PRESENTATION

B. Harrison: You indicated that both coat protein and the 3a gene play important parts in determining ability to infect plants systemically. Have you any indications as to whether either or both are needed for cell-to-cell movement of virus in inoculated leaves?

P. Ahlquist: We have done only a few experiments on local spread and so I cannot be dogmatic but it seems likely the situation will turn out to be somewhat similar to that for TMV, with some local spread being possible in the absence of coat protein.

J. Hammond: In the 3a protein of BMV mutant M2, which contains a variant form of BMV RNA3 and can infect that particular cowpea line, there are four amino acid changes. Are any of these predicted to be on the surface of the molecule?

P. Ahlquist: There are no physical data on the 3a protein structure so I could only speculate. Of those four amino acid changes, two make the protein more CCMV-like and two make it less CCMV-like, so the effect they will have is not clear. It is also entirely possible that the effective changes are in the non-coding regions and we need to do the additional genetics to resolve this point.

O. Barnett: Have you any idea how host specificity is determined?

P. Ahlquist: Again it would just be speculation but one could make simple models which would be consistent with the situation. For example, there could be a requirement for specificity in the 3a gene and the alteration in this gene in the M2 strain might allow it to move systemically in this particular cowpea line, with the cowpea line contributing a more relaxed tolerance of the BMV 1a and 2a genes. The situation becomes somewhat complex when we look at different combinations of RNA1, RNA2 and RNA3 in different hosts. Some of the early data suggested that functions on each of these genome parts are largely independent of one another but some more recent experiments suggest that just as we are seeing requirements for compatability between RNA1 and RNA2, so too there may be requirements for compatability between all three RNA species and between these and host component(s). This is not particularly consistent with some of the gating models of cell to cell movement that have been considered. However, the data are not sufficient to make any firm pronouncements.

Genetic Mapping of Cucumber Mosaic Virus

Michael Shintaku and Peter Palukaitis

Department of Plant Pathology, Cornell University, Ithaca, NY 14853, USA

Cucumber mosaic virus (CMV) is an isometric plant virus with a broad host range and a functionally divided genome (Peden and Symons, 1973; Kaper and Waterworth, 1981). RNAs 1 (3.3 kb) and 2 (3.0 kb) each encode one protein (111 kD and 97 kD, respectively), both of which are essential for virus replication (Nitta *et al.*, 1988). RNA 3 (2.2 kb) encodes the viral coat protein (24 kD) and an additional protein (30 kD) thought to be involved in potentiating the cell-to-cell movement of the virus (Davies and Symons, 1988).

Various strains of CMV exhibit differences in pathogenicity, host range, rates of replication, effects of temperature on replication, efficiency of replication of the satellite RNA of CMV, seed transmissibility and transmission efficiencies by the aphid vectors of CMV (Hitchborn, 1956; Mossop and Francki, 1977; Rao and Francki, 1982; Banik *et al.*, 1983; Edwards *et al.*, 1983; Lakshman and Gonsalves, 1985; Davis and Hampton, 1986; M. Roossinck and P. Palukaitis, unpublished data). A number of these properties have been delimited to specific RNAs by reassortment of the genomes of selected strains, a process called pseudorecombination (Mossop and Francki, 1977; Rao and Francki, 1982; Edwards *et al.*, 1983; Lakshman and Gonsalves, 1985; M. Roossinck and P. Palukaitis, unpublished data). However, until recently, it was not possible to further delimit the sequences involved in the host-virus interactions leading to the expression of the various properties.

We have constructed full-length cDNA clones of the three RNAs of the Fny-strain of CMV and generated infectious transcripts from these cDNA clones (Rizzo, 1989). We have also generated infectious RNA transcripts from a full-length cDNA clone of RNA 3 of the M-strain of CMV, which has a number of different biological properties. In this report, we describe the use of pseudorecombinants (Fny-CMV RNAs 1 plus 2 and M-CMV RNA 3) and recombinants between (cDNA clones of) Fny-CMV RNA 3 and M-CMV RNA 3 to delimit specific regions involved in the elicitation of host responses. Nucleotide sequence comparisons of Fny-CMV RNA 3 and M-CMV RNA 3 indicate that only a few nucleotide changes are involved in the alteration of the host response to virus infection.

PROPERTIES OF TWO STRAINS OF CMV

Previous work by Francki and colleagues on the M-strain of CMV (Mossop *et al.*, 1976; Mossop and Francki, 1977; Rao and Francki, 1982) and in this laboratory, on the Fny-strain of CMV has shown that these two strains of CMV exhibit a number of different properties (Table 1).

Table 1. Properties of two strains of CMV.

Property	Fny-CMV	M-CM	Delimited to RNA[a]	Reference[a]
Aphid-transmissible	yes	no	3	Mossop and Francki (1977)
Symptoms in tobacco	light-green/ dark-green mosaic	yellow chlorosis	2 or 3	Rao and Francki (1982)
Virus purification procedure	standard	novel	3	Mossop *et al* (1976)
Infection of maize	yes	no	2	Rao and Francki (1982)
Infection of squash	yes	no	3	This report

[a] The property associated with M-CMV RNA delimited to a particular RNA by pseudorecombination with other strains of CMV, indicated in the given citation.

Several of these properties, such as the inability to be transmitted by *Myzus persicae* and *Aphis gossypii*, the induction of systemic chlorosis in tobacco, an altered virus structure necessitating a novel CMV purification procedure, and an inability to infect squash (*Cucurbita pepo* - see below), all map to RNA 3 of M-CMV. Thus, the nucleotide sequence analysis of M-CMV RNA 3 and Fny-CMV RNA 3 was undertaken to determine the extent, nature and loci of sequence dissimilarities. Since there were 24 nucleotide (nt) differences between M-CMV and Fny-CMV in the 2216 nt of RNA 3, it was clear that relatively few changes could considerably alter various properties of a virus. However, since the nucleotide sequence changes were dispersed throughout the RNA, it was not possible to relate specific sequence alterations to the various differences in properties.

DELIMITATION OF INTERACTIVE DOMAINS BY RECOMBINATION

To further delimit the nucleotide sequence domains involved in eliciting chlorosis in tobacco and resistance to infection in squash, a series of recombinants was constructed between cDNA clones of Fny-CMV RNA 3 and M-CMV RNA 3 (Fig. 1). Plants were then inoculated with RNA transcribed by T7 RNA polymerase from the cDNA clones of Fny-CMV RNAs 1 and 2 and the recombinant cDNA clones of RNA 3.

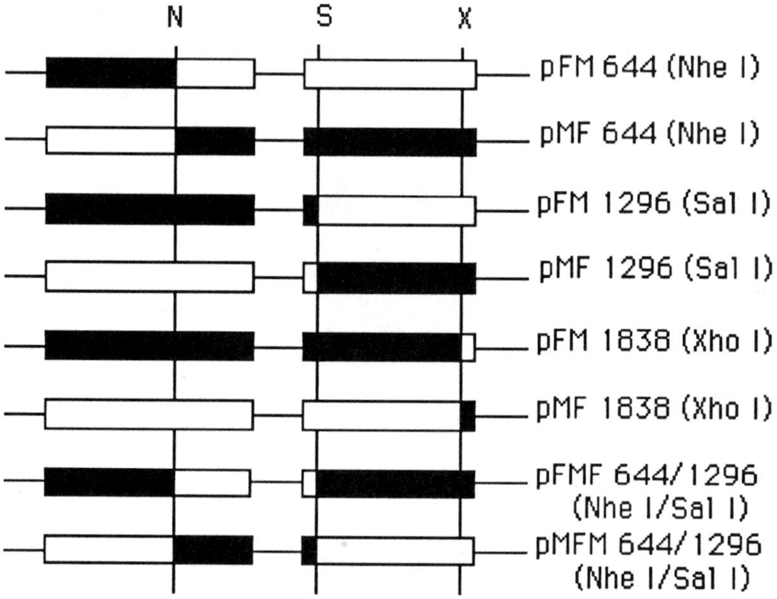

Fig. 1 Recombinant constructs between the cDNA clones of Fny-CMV RNA 3 and M-CMV RNA 3. Recombination occured at the specified restriction endonuclease sites. Symbols: N=Nhe I; S=Sal I; X=Xho I; ■ =Fny-CMV cDNA; □ =M-CMV cDNA. The plasmid constructs are named after the component, parental virus, cDNA clones and the site of recombination. Rectangular boxes represent open reading frames.

Table 2 shows that the systemic yellow chlorosis is induced by the pseudorecombinant Fny-CMV RNAs 1 plus 2 and M-CMV RNA 3, as well as by recombinants that contain most of the coat protein gene of M-CMV; i.e., nucleotides 1296-1838. The coat protein gene of M-CMV is located between nucleotides 1257 and 1910, and there are eleven differences in the nucleotide sequence and seven differences in the amino acid sequence between nucleotides 1296-1838 (Fig. 2).

Table 2. Symptoms induced by infectious transcripts of recombinant cDNA clones of CMV on tobacco.

Transcripts	Symptoms
F1F2F3 (parental)[a]	green mosaic
F1F2M3 (pseudorecombinant)[a]	yellow chlorosis
FM644 (Nhe I)[b]	yellow chlorosis
MF644 (Nhe I)[b]	green mosaic
FM1296 (Sal I)[b]	yellow chlorosis
MF1296 (Sal I)[b]	green mosaic
FM1838 (Xho I)[b]	green mosaic
MF1838 (Xho I)[b]	yellow chlorosis
FMF644/1296 (Nhe I/Sal I)[c]	green mosaic
MFM644/1296 (Nhe I/Sal I)[c]	yellow chlorosis

[a] F1F2F3 and F1F2M3 refer to T7 RNA transcripts synthesized from non-recombinant constructs.
[b] Recombinants were constructed between cDNA cones of Fny-CMV RNA 3 and M-CMV RNA 3. Plants were inoculated with Fny-CMV RNA 1 plus 2 and an RNA transcript of the recombinant constructs. Thus, FM644 contains sequences of Fny-CMV RNA 3 from the 5'-end to nucleotide 644, the Nhe I site, and M-CMV RNA 3 sequences from nucleotide 644 to the 3'-end (nucleotide 2216). MF644 is the reciprocal of FM644.
[c] These are transcripts of recombinants involving exchanges of the Nhe I-Sal I fragment.

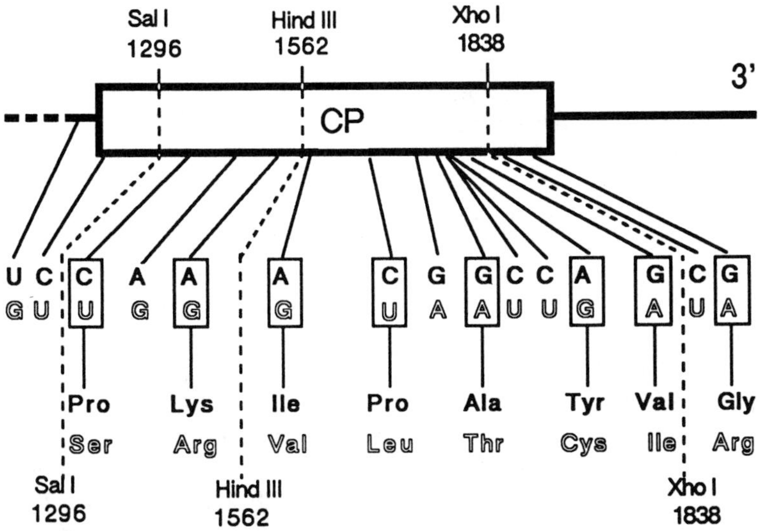

Fig. 2 Diagram of the coat protein (CP) gene of CMV. The differences in the nucleic acid and amino acid sequence in the 3' halves of Fny-CMV RNA 3 (bold letters) and M-CMV RNA 3 (hollow letters) are indicated, as are the positions of several restriction endonuclease cleavage sites.

The M-strain of CMV does not infect squash (*C. pepo*) (Table 3). By using pseudorecombinants between M-CMV and other CMV strains capable of infecting squash, the inability to infect squash can be delimited to RNA 3 of M-CMV (Table 3). Thus, the pseudorecombinants and recombinants constructed in this study were used to further delimit this property.

Table 3. Infectivity of CMV strains and pseudorecombinants in *Cucurbita pepo* (squash).

Strain of CMV[a]	Infectivity in squash
G-CMV	+
M-CMV	-
Q-CMV	+
Fny-CMV	+
G1G2M3-CMV[b]	-
M1M2G3-CMV	+
M1M2Q3-CMV	+
Q1Q2M3-CMV	-
F1F2M3-CMV	-

[a] Strains G-, M-, Q- and pseudorecombinats G1G2M3, M1M2G3, M1M2Q3, and Q1Q2M3 were obtained from R. I. B. Francki, Adelaide, Australia.
[b] Pseudorecombinats are indicated by their composition; e.g., F1F2M3 is RNA 1 and 2 of the Fny-strain and RNA 3 of the M-strain.

Buffered sap from the tobacco plants inoculated with the various transcripts was applied to squash cotyledons. The results of this infectivity assay are shown in Table 4. Once again, it seems that the determinant for elicitation of resistance in squash is associated with the coat protein; however, other recombinants between the cDNA clones of M-CMV RNA 3 and Fny-CMV RNA 3 need to be tested to confirm this observation. Moreover, the effects of various sequences flanking nucleotides 1296-1838 on the elicitation of the resistance response also need to be analyzed, since transcripts of one of the recombinants containing only two nucleotide (one amino acid) alterations of the M-CMV coat protein sequence [transcript MF1838 (Xho I); see Table 4] was able to induce a systemic infection. This infection was not apparent at 6 days post-inoculation, by which time plants inoculated with transcripts of all of the recombinants containing the coat protein gene of Fny-CMV showed systemic systems; however, by twelve days post-inoculation a mild systemic infection, in the form of sporadic, chlorotic, local lesions, was evident on plants inoculated with transcript MF1838. Over the next few weeks, these plants continued to only show signs of a mild infection. Replication of CMV was not detectable by dot-blot hybridization in the symptomless plants (results not presented).

Table 4. Infection of *Cucurbita pepo* by infectious transcripts of recombinant cDNA clones of CMV.

Transcripts	Symptoms
F1F2F3	stunting, epinasty, severe mosaic
F1F2M3	no infection
FM644 (Nhe I)[a]	no infection
MF644 (Nhe I)	stunting, epinasty, severe mosaic
FM1296 (Sal I)	no infection
MF1296 (Sal I)	stunting, epinasty, severe mosaic
FM1838 (Xho I)	stunting, epinasty, severe mosaic
MF1838 (Xho I)	mild infection, sporadic chlorotic lesions

[a] Recombinant transcripts as described in Table 2.

Both the lack of aphid-transmissibility associated with M-CMV RNA 3 and the requirement for an altered virus purification protocol, presumably reside in the coat protein gene. Thus, all of the characteristics ascribed to RNA 3 of M-CMV (Table 1) are localized in the coat protein gene.

The coat protein gene needs to be further bifurcated to delimit domains involved in the above interactions. Recombinations at the Hind III site at nucleotide 1562 (see Fig. 2) would split the coat protein such that two of the eight amino acid differences between M-CMV and Fny-CMV would be in the N-terminal-half of coat protein, and five of the amino acid differences would be between the Hind III site and the Xho I site (Fig. 2). The construction of the two Hind III recombinants is in progress.

DISCUSSION

The M-strain of CMV is a mutant of Price's No. 6 strain of CMV (P6-CMV; Price, 1934; see Mossop *et al.*, 1976). The latter strain induces systemic chlorosis in tobacco, is aphid-transmissible and is purified by the standard extraction procedure (Roberts and Wood, 1981). Recently, a variant of M-CMV was isolated that had regained its aphid-transmissibility (R. Francki, personal communication). Thus, the altered virus structure necessitating a novel purification procedure, the chlorosis induction and the aphid-transmissibility are associated with independent determinants in the viral coat protein gene. Assuming that all of these determinants are associated with alterations in the viral coat protein itself, rather than in the nucleotide sequence of the coat protein gene, then eight or less changes in amino acid sequence (Fig. 2) can alter the above three properties of CMV as well as the ability of CMV to infect squash.

The ability of an altered, but structurally functional, viral coat protein to induce resistance to viral infection in one host is an unexpected finding. This suggests either that a host can restrict viral infection in response to the presence of elicitors in the viral coat protein [as in the induction of local lesions in tobacco containing the N'-gene by the coat protein of tobacco mosaic virus (Saito *et al.*, 1987; Knorr and Dawson, 1988; Culver and Dawson, 1989)], or that viral coat protein may have some function other than encapsidation that could be interfered with in a given host species [as shown for other coat protein mutants of tobacco mosaic virus (Dawson *et al.*, 1988)]. Since many "squash cultivars" are actually different species (*e.g., Cucumis melo, Cucurbita moschata*, or *Cucurbita pepo*), experiments also need to be done to determine the cultivar/species specificity of this phenomenon.

By comparing the nucleotide or amino acid sequence of the coat protein genes of subgroup I CMV strains, it might be possible to further delimit what sequence changes are unique to M-CMV. However, this requires making three assumptions that cannot be substantiated: (i) that the nucleotide sequences determined for CMV strains C, D, O, and Y accurately reflect the nucleotide sequence of these strains at the time they were characterized biologically (e.g., are all of these strains still aphid-transmissible after years of mechanical passage?); (ii) that a particular property is always determined by the same nucleotide sequence changes rather than alterations at one of several loci; (iii) that nucleotide sequence changes at other loci do not affect the expression of a particular property; i.e., we cannot rule out the effect of either variability in the genetic background or multiple determinants acting synergistically on the expression of the various properties. Clearly, additional recombinations and site-directed mutagenesis of the various nucleotides differing in this region of the two coat proteins will be instrumental in localizing the determinants for aphid-transmissibility, chlorosis-induction, altered virus structure and alteration in the host range *vis-à-vis* squash.

ACKNOWLEDGEMENTS

This work was supported by Grant DE-FG02-ER13505 from the Department of Energy and Grant No. 88-37263-3806 from the USDA CGO. Michael Shintaku was supported in part by a fellowship from the Cornell University Plant Science Center.

REFERENCES

Banik, M., Zitter, T. A., and Lyons, M. E. (1983). A difference in virus titer of two cucumber mosaic virus isolates as measured by ELISA and aphid transmission. *Phytopathology* **73**, 362 (Abstr.).

Culver, J. N., and Dawson, W. O. (1989). Point mutations in the coat protein gene of tobacco mosaic virus induce hypersensitivity in *Nicotiana sylvestris*. *Mol. Plant-Microbe Interact.* **2**, 209-213.

Davies, C. and Symons, R. H. (1988). Further implications for the evolutionary relationships between tripartite plant viruses based on cucumber mosaic virus RNA 3. *Virology* **165**, 216-224.

Davis, R. F., and Hampton, R. O. (1986). Cucumber mosaic virus isolates seedborne in *Phaseolus vulgaris*: Serology, host-pathogen relationships and seed transmission. *Phytopathology* **76**, 999-1004.

Dawson, W. O., Bubrick, P., and Grantham, G. L. (1988). Modifications of the tobacco mosaic virus coat protein gene affecting replication, movement and symptomatology. *Phytopathology* **78**, 783-789.

Edwards, M. C., Gonsalves, D., and Provvidenti, R. (1983). Genetic analysis of cucumber mosaic virus in relation to host resistance: Location of determinants for pathogenicity to certain legumes and *Lactuca saligna*. *Phytopathology* **73**, 269-273.

Hitchborn, J. H. (1956). The effects of temperature on infection with strains of cucumber mosaic virus. *Ann. Appl. Biol.* **44**, 590-598.

Kaper, J. M., and Waterworth, H. E. (1981). Cucumoviruses. *In* "Handbook of Plant Virus Infections and Comparative Diagnosis" (E. Kurstak, Ed.), pp. 257-332. Elsevier/Holland Biomedical., New York.

Knorr, D. A., and Dawson, W. O. (1988). A point mutation in the tobacco mosaic virus capsid protein gene induces hypersensitivity in *Nicotiana sylvestris*. *Proc. Natl. Acad. Sci. USA* **85**, 170-174.

Lakshman, D. K., and Gonsalves, D. (1985). Genetic analysis of the large lesion mutants of two cucumber mosaic virus strains. *Phytopathology* **75**, 758-762.

Mossop, D. W., and Francki, R. I. B. (1977). Association of RNA 3 with aphid transmission of cucumber mosaic virus. *Virology* **81**, 177-181.

Nitta, N., Takanami, Y., Kuwata, S., and Kubo, S. (1988). Inoculation with RNAs 1 and 2 of cucumber mosaic virus induces viral RNA replicase activity in tobacco mesophyll protoplasts. *J. Gen. Virol.* **69**, 2695-2700.

Peden, K. W. C., and Symons, R. H. (1973). Cucumber mosaic virus contains a functionally divided genome. *Virology* **53**, 487-492.

Price, W. C. (1934). Isolation and study of some yellow strains of cucumber mosaic. *Phytopathology* **24**, 743-761.

Rao, A. L. N., and Francki, R. I. B. (1982). Distribution of determinants for symptom production and host range on the three RNA components of cucumber mosaic virus. *J. Gen. Virol.* **61**, 197-205.

Rizzo, T. M. (1989). Molecular analysis of RNAs 1 and 2 from cucumber mosaic virus and *in vitro* generation of infectious transcripts. PhD Thesis, Cornell University, 102 pp.

Roberts, P. L., and Wood, K. R. (1981). Protein and RNA composition of a mild (W) and a severe (P6) strain of cucumber mosaic virus. *Microbios* **32**, 37-46.

Saito, T., Meshi, T., Takamatsu, N., and Okada, Y. (1987). Coat protein gene sequence of tobacco mosaic virus encodes a host response determinant. *Proc. Natl. Acad. Sci., USA* **84**, 6074-6077.

DISCUSSION OF P. PALUKAITIS' PRESENTATION

T. Pirone: When you say the M strain is unable to infect squash, do you mean unable to infect it systemically or unable to infect even the inoculated leaves?

P. Palukaitis: No local lesions were produced in the inoculated cotyledons. On the one occasion when we obtained a systemic infection the inoculum consisted of purified virus at the very high concentration of 4 mg/ml: in this test one out of the nine plants showed a few lesions on a non-inoculated upper leaf. When virus was transferred from that leaf to tobacco, the typical yellowing symptoms were produced. When it was transferred back to squash, there was no systemic infection, so there was no evidence for selection of a resistance-breaking strain. Again no symptoms developed in the cotyledons but no tests were made in any of the experiments for virus production in the cotyledons.

B. Harrison: Is the result the same if you use RNA inocula?

P. Palukaitis: Yes. Inocula of virus RNA or virus particles produce the same result.

B. Raccah: Years ago we did some transcapsidation experiments with the yellow and mosaic strains of CMV, and found the yellow strain could then be transmitted by aphids. Did you test your pseudo-recombinants to check if the coat protein of the Fny strain can enable isolates containing the RNA1 and RNA2 of the M strain to be transmitted by aphids?

P. Palukaitis: We have not done those experiments but Francki has found that when RNA of the M strain or of tomato aspermy virus is encapsidated in the coat protein of other strains of CMV the viruses become transmissible by aphids. These experiments were done both by re-assembly of virus particles and by producing pseudo-recombinants.

Mutational Analysis of Viroid Pathogenicity and Movement

Robert A. Owens and Rosemarie W. Hammond

Microbiology and Plant Pathology Laboratory, Plant Sciences Institute, USDA-ARS Beltsville, MD 20705, USA

Viroids are the smallest known agents of infectious disease--small (246 to 375 nucleotides), single stranded, circular RNA molecules that are unencapsidated and lack detectable messenger activity. Viroids replicate autonomously, and the absence of viroid-specified proteins implies that replication and disease induction must result from their ability to interact directly with certain host constituents (reviewed in Diener, 1987). These characteristics, together with the existence of extensive nucleotide sequence data, make viroids uniquely suited for molecular studies of host-pathogen interaction.

Comparative sequence analysis suggests that viroids contain five structural domains (Keese & Symons, 1985), and the conformation of one of these domains, the pathogenicity domain in the left side of the rod-like native structure, appears to control symptom expression in potato spindle tuber viroid (PSTV) and related viroids. Visvader & Symons (1986) have shown that symptom production by novel citrus exocortis viroid (CEV) chimeras is controlled by the source of the pathogenicity domain. In these studies, recombinant DNA techniques were used to construct full-length infectious cDNAs from fragments derived from naturally-occuring mild and severe sequence variants. Schnölzer *et al.*, (1985) have suggested that the thermal stability of a "virulence modulating" region within the pathogenicity domain of PSTV controls symptom expression by regulating its ability to interact with unidentified host factor(s). However, symptom severity cannot always be correlated with the calculated stability of the pathogenicity domain, as was shown for CEV (Visvader & Symons, 1985).

Pathogenicity is only one of the many interesting biological properties of viroids, but it is uniquely accessible to experimental investigation. The infectivity of either longer-than-full-length viroid cDNAs (Cress *et al.*, 1983) or their respective RNA transcripts (Ohno *et al.*, 1983) allows the use of site-directed mutagenesis techniques to identify structural features responsible for the ability of viroids to replicate, move systemically, and cause disease. In a recently completed series of studies we have determined the effect of single and multiple nucleotide substitutions upon symptom expression by PSTV and tomato apical stunt viroid (TASV). Some (but not all) of the effects observed are consistent with the properties of naturally-occuring variants. One particularly interesting PSTV variant exhibited a dramatically decreased ability to spread systemically in tomato.

ISOLATION OF VIROID VARIANTS

Random and site-directed mutagenesis of PSTV

Figure 1 shows the 359 nucleotides of PSTV (Intermediate) arranged in its familar rod-like and highly base-paired native structure. TASV, a second viroid that has also been subjected to mutational analysis, contains 360 nucleotides and exhibits 64% overall sequence similarity with PSTV.

We have previously shown that multiple nucleotide substitutions within the left (PSTV-P) and right (PSTV-R) terminal loops of PSTV abolish the infectivity of RNA transcripts containing an 11 nucleotide (GGATCCCCGGG) sequence duplication derived from the upper portion of the central conserved region (Hammond & Owens, 1987). Although conversion of a G:C base-pair within the central conserved region to an A:U pair had no effect upon specific infectivity, replication and/or systemic spread of the resulting variant (PSTV-H+H') was impaired.

We subsequently used the procedures described by Myers *et al.*, (1985a) to introduce random single-base substitutions into PSTV by treatment of single-stranded PSTV cDNAs with nitrous acid followed by synthesis of the complementary strand with reverse transcriptase. The resulting duplex DNA fragments containing random single-base substitutions were cloned, amplified as a population, and separated from wild-type DNA by either denaturing gradient (Myers *et al.*, 1985a) or temperature gradient (Riesner *et al.*, 1989) gel electrophoresis. Bioassay of 118 randomly selected PSTV cDNA clones on tomato seedlings (cv. Rutgers) revealed that 58 were infectious.

Nineteen of those 58 infectious cDNAs (33%) could be distinguished from wild-type PSTV cDNA by temperature gradient gel electrophoresis, strongly suggesting the presence of one or more nucleotide substitutions in the left side of the native structure (G. Steger, personal communication). Nucleotide sequence analysis of 18 infectious cDNAs revealed that we had isolated 10 different altered cDNAs: 6 with single nucleotide substitutions, 2 with three substitutions, and 1 each with either two or five substitutions. All 10 cDNAs contained at least one substitution in the left side of their native structure, and nearly 50% of all substitutions (8 of 18) were located in the pathogenicity domain (see Fig. 1). A somewhat higher percentage of the noninfectious cDNAs (48%) exhibited altered electrophoretic mobilities, and determination of their nucleotide sequences is currently underway.

Thus far, progeny arising from 5 different mutant cDNAs have been characterized by nucleotide sequence analysis of PSTV cDNA amplified by the polymerase chain reaction (PCR). Substitutions at positions 310, 311, 340, 1, and 13 were retained in the progeny, but the A->G substitution at position 60

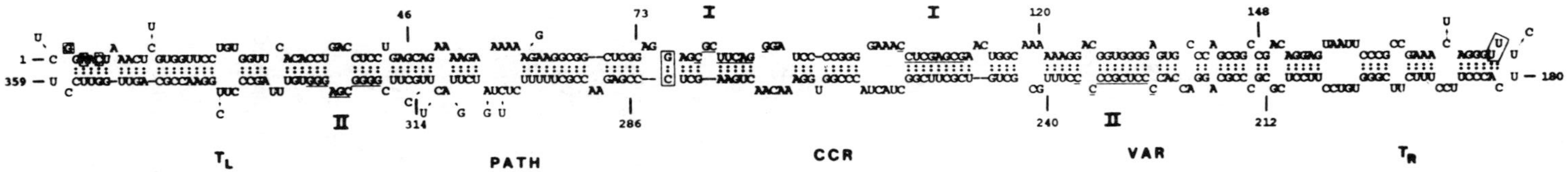

Fig. 1 Native structure of PSTV (Intermediate). Locations of secondary hairpins I and II (underlined) and the boundaries of the left (T_L) and right (T_R) terminal loops, pathogenicity (PATH), central conserved (CCR), and variable (VAR) domains are marked. Nucleotide substitutions described by Hammond & Owens (1987) have been boxed, while those introduced by random chemical mutagenesis are indicated by arrows.

was unstable and reverted to the wild-type sequence. The C->U substitution at position 311 is of particular interest because, as shown in Fig. 2, it should make the secondary structure of the pathogenicity domain very similar to that of a naturally-occuring mild strain. The A->G substitution at position 310, on the other hand, should leave the structure of the pathogenicity domain almost unaltered. Both these variants induced mild symptoms in Rutgers tomato seedlings (Owens *et al.*, in preparation). The substitutions at positions 303, 305, and 312 should also have significant structural effects, but their progeny have not yet been characterized.

In a second series of experiments, as many as four nucleotide substitutions were introduced into the pathogenicity domain of PSTV (Intermediate) by oligonucleotide-directed mutagenesis (Hammond, in preparation). As shown in Fig. 2, the presence of all four substitutions converts the intermediate strain into the more severe KF-440 strain (Schnölzer *et al.*, 1985). Only two cDNAs containing fewer than four substitutions were infectious, i.e., constructs M1 and M2/M1. Both variants induce milder symptoms in tomato than the parental intermediate strain, and nucleotide sequence analysis of the progeny proved that the alterations in both M1 and M2/M1 are stably maintained.

Saturation mutagenesis within the pathogenicity domain of TASV

Two naturally-occuring strains of TASV have been described. Very similar in length (360 vs. 363 nucleotides) and exhibiting extensive (91%) sequence similarity, the Ivory Coast and Indonesian isolates of TASV both induce severe symptoms in tomato (Kiefer *et al.*, 1983; Candresse *et al.*, 1987). As shown in Fig. 3, natural sequence variation within the pathogenicity domain is confined to the predominantly helical region adjoining the central conserved region. Indeed, the compensatory nature of many of these changes strongly supports the base-pairings shown.

When a series of 12 single nucleotide (A->G) substitutions between positions 48 and 76 in the TASV pathogenicity domain were tested for their effect upon infectivity and symptom expression, not one of these mutations proved to be lethal. Several infectivity trials were conducted using both recombinant plasmid DNAs and the respective T3 RNA polymerase transcripts as inocula, and all 12 mutant cDNAs gave rise to progeny in at least one trial. The resulting TASV progeny were characterized by RNase mapping (Myers *et al.*, 1985a) and nucleotide sequence analysis of PCR-amplified cDNAs. As shown in Fig. 3, three of the 12 substitutions tested gave rise to altered progeny. The patterns of sequence changes observed were unexpectedly complex, however.

Introduction of an A->G substitution at position 61 produced a mixture of variants whose pathogenicity domains resemble that of PSTV, i.e. a small U-containing interior loop has been incorporated into the predominantly helical region adjoining the central conserved region. Rearrangements initiated by the

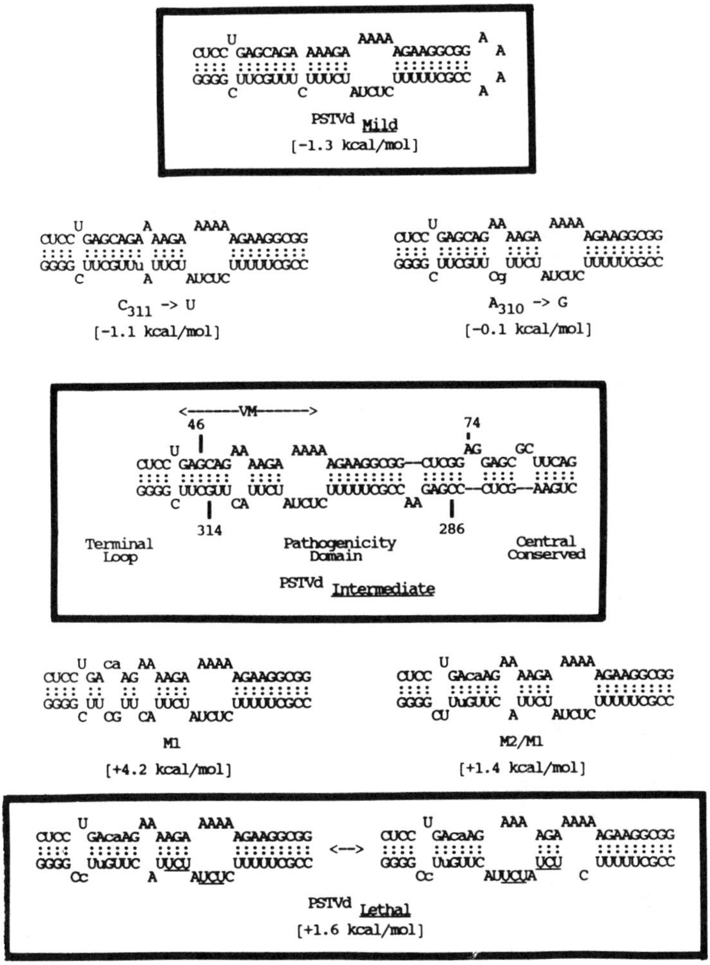

Fig. 2 PSTV variants containing mutations within the pathogenicity domain. Variants C311/U and A310/G were isolated after nitrous acid mutagenesis of single-stranded PSTV cDNA, while M1 and M2/M1 were created by oligonucleotide-directed mutagenesis. Secondary structures for the naturally-occuring mild, intermediate, and lethal strains have been redrawn from Schnölzer *et al.* (1985) who discuss the possible significance of the virulence-modulating (VM) region and the alternative structures shown for PSTV (Lethal). Lowest free energy structures for the variants were calculated by applying the algorithm of Zuker (1989) to model oligonucleotides analogous to that shown for PSTV (Mild). Although slightly different structures for the virulence-modulating region were obtained when different model oligonucleotides were used for these calculations, the relative effects of the sequence alterations are the same.

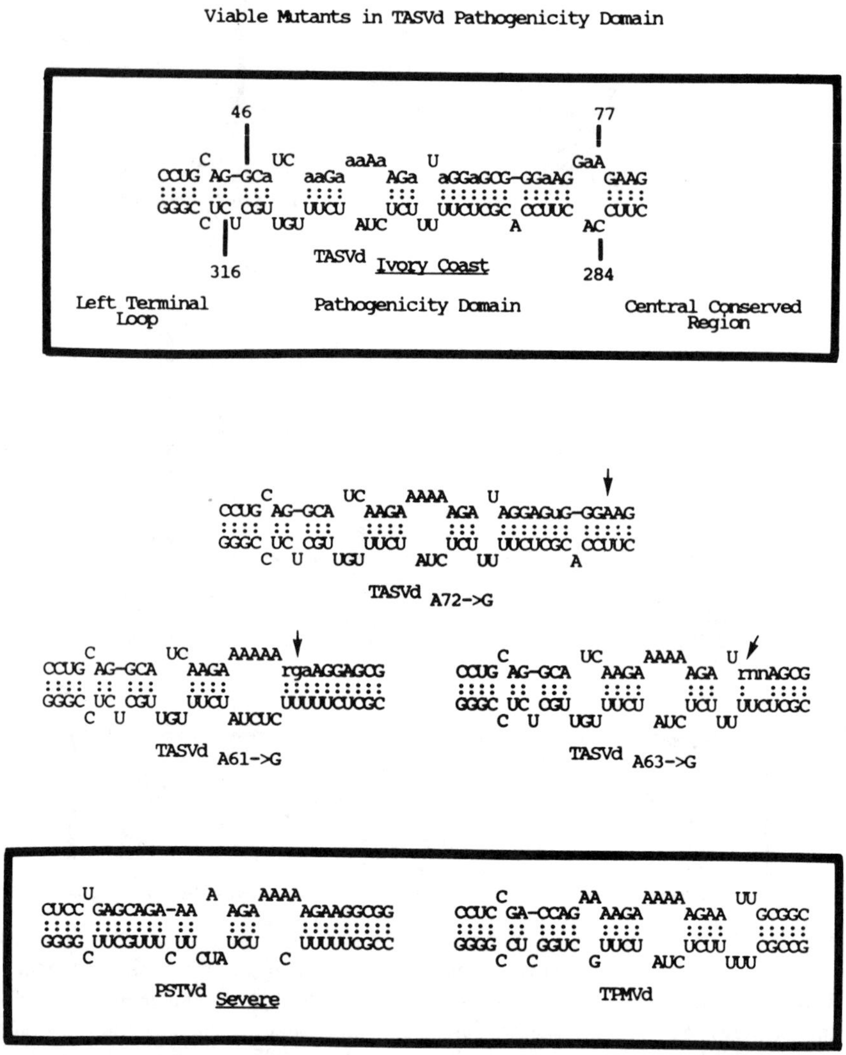

Fig. 3 TASV variants containing alterations in the pathogenicity domain. Positions of the 12 initial A->G substitutions and natural sequence variation within the pathogenicity domain of TASV have been marked. For the three variants, the positions of the initial A->G substitutions at positions 61, 63, and 72 are indicated by arrows, while the sequence alterations present in the progeny are denoted by lower case letters (r, purine; n, mixture of A,C,G, and U). Secondary structures for the pathogenicity domains of PSTV (Severe) and TPMV have been included for purposes of comparison.

substitution at position 63 may yield a pathogenicity domain whose structure resembles that of tomato planta macho viroid (TPMV). When tomatoes were inoculated with TASV RNA transcripts containing an A->G substitution at position 72, the progeny which subsequently appeared contained a C->U substitution at position 68. The initial A->G substitution at position 72 appears to have reverted to wild-type. All three variants or variant mixtures induce symptoms that are indistinguishable from those of the parental isolate, TASV (Ivory Coast).

UNCOUPLING REPLICATION FROM CELL-TO-CELL MOVEMENT

We have previously reported that the introduction of multiple nucleotide substitutions into either the left or right terminal loop of PSTV abolishes cDNA infectivity (Hammond & Owens, 1987; see Fig. 1). In subsequent experiments, these mutations (PSTV-P and PSTV-R, respectively) were used to demonstrate that multimeric viroid RNAs can be processed at sites other than the upper portion of the central conserved region to yield monomeric progeny (Hammond *et al.*, 1989). Limitations inherent in the systemic bioassay, however, left us uncertain as to whether or not these mutations (and especially PSTV-R) were actually lethal. Perhaps their ability to replicate and/or move systemically has been severely restricted but not abolished. Such questions are pertinent because *single* substitutions in either loop were not lethal (see Fig. 1).

To answer these questions, a series of infectivity studies in which viroid inoculation was mediated by the Ti plasmid of *Agrobacterium tumefaciens* were carried out (Gardner *et al.*, 1986; Hammond, in preparation). Full-length wild-type and PSTV-R cDNAs were placed under the control of the CaMV 35S promoter to maximize transcription *in vivo,* and both cDNAs contained an 11-nucleotide terminal repeat (GGATCCCCGGG) to maximize the specific infectivity of the transcripts. In addition to dot-blot hybridization of sap samples prepared from uninoculated upper leaves, a combination of Northern analysis (Salazar *et al.*, 1988) and *in situ* hybridization (Barker *et al.*, 1988) was used to determine the distribution of PSTV-related RNAs in inoculated tomato seedlings.

Dot-blot and Northern analyses of RNA extracted from galls, roots, and leaves harvested from plants inoculated with wild-type PSTV cDNA revealed that viroid replication was systemic, i.e. both circular progeny and double-stranded replicative intermediates were present in all tissues tested. Parallel analyses of plants inoculated with *A. tumefaciens* containing PSTV-R cDNA showed viroid replication to be localized within the galls and roots; no viroid-related RNA could be detected in the foliage. Nucleotide sequence analysis of the PSTV-R progeny showed that both U->A substitutions had been retained. Examination of tissue collected from the lower portions of the stems or roots by *in situ* hybridization using an 35S-labeled probe for PSTV (+) strands showed that, in contrast to wild-type PSTV, the mutant was confined to the phloem tissue. PSTV-R progeny purified from infected gall or root tissue was also used to inoculate tomato cotyledons in a conventional bioassay. Unlike the linear PSTV-R transcripts previously tested (see Hammond & Owens, 1987), circular PSTV-R seemed able to replicate and spread systemically with near-normal kinetics.

CONCLUSIONS AND REMAINING QUESTIONS

Contrary to initial indications (e.g. Ishikawa *et al.*, 1985; Hammond & Owens, 1987), viroids appear able to undergo significant structural alteration without losing their ability to replicate and move systemically. Most alterations within regions believed important for replication (e.g., the upper portion of the central conserved region or secondary hairpin II, see Fig. 1) abolish infectivity, but viable mutations in several other regions have been isolated and characterized. For PSTV and TASV, the pathogenicity domain appears to be particularly tolerant, and several variants with interesting biological properties have been isolated. Nonlethal alterations appear to be clustered in areas which exhibit natural sequence variation. Given this ability to isolate viable viroid mutants, what role can such mutants play in bringing new techniques/concepts to bear upon such perennial questions as: What are the molecular mechanisms responsible for viroid replication? How do viroids interact with their host to cause disease? What determines viroid host range?

With respect to replication mechanisms, *in vitro* mutagenesis has highlighted the importance of the upper portion of the central conserved region for cDNA infectivity (Visvader *et al.*, 1985; Hammond & Owens, 1987), and this result has been confirmed by studies of the conversion of multimeric PSTV RNA transcripts into circular monomers by nuclear extracts (Tsagris *et al.*, 1987). Mutagenesis has also been used to detect the presence of alternative sites for the *in vivo* processing of multimeric viroid RNAs (Hammond *et al.*, 1989). The full power of site-directed mutagenesis techniques will only be realized, however, when improved *in vitro* assays for viroid replication/processing are developed (see Sheldon & Symons, 1989, for an example involving a self-cleaving satellite RNA).

On the other hand, mutagenesis has already provided new insights into structural features responsible for pathogenicity and host range. Recent speculation about the mechanism of viroid pathogenesis has centered upon the thermodynamic stability of a virulence-modulating region within the PSTV pathogenicity domain (Schnölzer *et al.*, 1985) and the ability of nearby nucleotides to base-pair with the 5'-terminus of a 7S host RNA that may be involved in protein translocation (Haas *et al.*, 1988). Several of our PSTV variants induce milder symptoms than the parental Intermediate strain, but, unlike in the case of naturally-occuring isolates, structural calculations suggest that only one mutation (i.e., the C->U substitution at position 311) stabilizes the pathogenicity domain. Even for PSTV, factors other than thermal stability of the virulence-modulating region appear to be important for symptom expression.

The three TASV variants which we have isolated all contain alterations in a different portion of the pathogenicity domain, the predominantly helical region which adjoins the central conserved region. These alterations do not affect symptom expression, and the potential structural rearrangements appear not to affect the conformation/stability of the virulence-modulating region. Thus, the properties of these mutants provide indirect support for the existence of such a region within the pathogenicity domain. To clarify the role of the virulence-modulating region in pathogenicity will require that we search for interactions of host proteins with the pathogenicity domain and that we determine experimentally (rather than calculate) its conformation *in vivo*.

We would also like to know more about the apparently complex pathway responsible for the origin/evolution of these TASV variants. At present, we can only speculate about the selective pressures that are present, but they may be host-specific. If so, host-specific differences in progeny sequence could provide needed insight into viroid-host interaction and structural features which determine host range. Although viral-encoded determinants of host range and virus-host interaction have been identified by the systematic construction of chimeras (e.g., Schoelz & Shepherd, 1988; Dawson *et al.*, 1988), a similar strategy does not appear to be possible for viroids.

Finally, if mutagenesis within the right terminal loop has indeed uncoupled PSTV replication from cell-to-cell movement, we foresee several exciting consequences for viroid research. The ability to spread from cell-to-cell is obviously an important determinant of viroid host range, but, more importantly, the ability to genetically isolate two normally interrelated biological processes greatly enhances our ability to study them at the molecular level. Newer analytical techniques (e.g., *in situ* hybridization) provide us with the ability to separate them spatially as well. Both capabilities are significant additions to the predominantly biochemical approach currently employed in viroid research.

ACKNOWLEDGEMENTS

We thank S.M. Thompson, M. Hale, and D.R. Smith for skilled technical assistance. These studies were partially supported by the USDA Competitive Research Grants Program (Grants 85-CRCR-1738 and 88-37263-3990).

REFERENCES

Barker, S.J., Harada, J.J., & Goldberg R.B. 1988. Cellular localization of soybean storage protein mRNA in transformed tobacco seeds. *Proc. Natl. Acad. Sci. USA* **85**, 458-462.

Candresse, T., Smith, D., & Diener, T.O. 1987. Nucleotide sequence of a full-length infectious clone of the Indonesian strain of tomato apical stunt viroid (TASV). *Nucleic Acids Res.* **15**, 10597.

Cress, D.E., Kiefer, M.C., & Owens R.A. 1983. Construction of infectious potato spindle tuber viroid cDNA clones. *Nucleic Acids Res.* **11**, 6821-6835.

Dawson, W.O., Knorr, D.A., and Bubrick, P. 1988. Elicitor active coat protein mutations in cDNA clones of tobacco mosaic virus. In: "Physiokogy and Biochemistry of Plant-Microbial Interactions" (N. Keen, T. Kosuge, and L.L. Walling, eds.), pp. 139-147. *American Society of Plant Physiologists.*

Diener, T.O. 1987. *The Viroids.* Plenum Press, New York.

Gardner, R.C., Chonoles K., & Owens R.A. 1986. Potato spindle tuber viroid infections mediated by the Ti plasmid of *Agrobacterium tumefaciens. Plant Mol. Biol.* **6**, 221-228.

Haas, B., Klanner, A., Ramm K., & Sanger, H.L. 1988. The 7S RNA from tomato leaf tissue resembles a signal recognition particle RNA and exhibits a remarkable sequence complementarity to viroids. *EMBO J.* **7**, 4063-4074.

Hammond, R.W., Diener, T.O., & Owens, R.A. 1989. Infectivity of chimeric viroid transcripts reveals the presence of alternative processing sites in potato spindle tuber viroid. *Virology* **170**, 486-495.

Hammond, R.W. & Owens, R.A. 1987. Mutational analysis of potato spindle tuber viroid reveals complex relationships between structure and infectivity. *Proc. Natl. Acad. Sci. USA*, **84**, 3967-3971.

Ishikawa, M., Meshi T., Okada, Y., Sano T., & Shikata E. 1985. *In vitro* mutagenesis of infectious viroid cDNA clone. *J. Biochem.* **98**, 1615-1620.

Keese P. & Symons, R.H. 1985. Domains in viroids: Evidence of intermolecular RNA rearrangements and their contribution to viroid evolution. *Proc. Natl. Acad. Sci. USA* **82**, 4582-4586.

Kiefer, M.C., Owens, R.A., & Diener, T.O. 1983. Structural similarities between viroids and transposable genetic elements. *Proc. Natl. Acad. Sci. USA* **80**, 6234-6238.

Kunkel T.A. 1987. Oligonucleotide-directed mutagenesis without phenotypic selection. In: *Current Protocols in Molecular Biology* (F.M. Ausubel et al., eds.). John Wiley & Sons, New York.

Myers, R.M., Larin, Z., & Maniatis T. 1985. Detection of single base substitutions by ribonuclease cleavage at mismatches in RNA:RNA duplexes. *Science* **230**, 1242-1246.

Myers, R.M., Lerman, L.S., and Maniatis T. 1985a. A general method for saturation mutagenesis of cloned DNA fragments. *Science* **229**, 242-247.

Ohno, T., Ishikawa, M., Takamatsu N., Meshi, T., Okada, Y., Sano T., & Shikata E. 1983. *In vitro* synthesis of infectious RNA molecules from cloned hop stunt viroid complementary DNA. *Proc. Japan Acad.* **59** (Series B), 251-254.

Owens, R.A., Candresse, T., & Diener, T.O. Construction of novel viroid chimeras containing portions of tomato apical stunt and citrus exocortis viroids. *Virology* (in press).

Riesner, D., Steger, G., Zimmat, R., Owens, R.A., Wagenhofer M., Hillen, W., Vollbach, S. & Henco, K. 1989. Temperature-gradient gel electrophoresis of nucleic acids: Analysis of conformational transitions, sequence variations, and protein-nucleic acid interactions. *Electrophoresis* **10**, 377-389.

Salazar, L.F., Hammond R.W., Diener T.O., & Owens, R.A. 1988. Analysis of viroid replication following *Agrobacterium*-mediated inoculation of non-host species with potato spindle tuber viroid cDNA. *J. Gen. Virol.* **69**, 879-889.

Schoelz, J.E. & Shepherd, R.J. 1988. Host range control of cauliflower mosaic virus. *Virology* **162**, 30-37.

Schnölzer, M., Haas B., Ramm, K., Hofmann, H., & Sänger, H.L. 1985. Correlation between structure and pathogenicity of potato spindle tuber viroid (PSTV). *EMBO J.* **4**, 2181-2190.

Sheldon, C.C. & Symons R.H. 1989. Mutagenesis analysis of a self- cleaving RNA. *Nucleic Acids Res.* **17**, 5679-5685.

Tsagris M., Tabler M., Mühlbach, H-P., & Sänger, H.L. 1987. Linear oligomeric potato spindle tuber viroid (PSTV) RNAs are accurately processed *in vitro* to the monomeric circular viroid proper when incubated with a nuclear extract from healthy potato cells. *EMBO J.* **6**, 2173-2183.

Visvader, J.E., Forster, A.C., & Symons R.H. 1985. Infectivity and in vitro mutagenesis of monomeric cDNA clones of citrus exocortis viroid indicates the site of processing of viroid precursors. *Nucleic Acids Res.* **13,** 5843-5856.

Visvader, J.E. & Symons, R.H. 1985. Eleven new sequence variants of citrus exocortis viroid and correlation of sequence with pathogenicity. *Nucleic Acids Res.* **13**, 5843-5856.

Visvader, J.E. & Symons R.H. 1986. Replication of in vitro constructed viroid mutants: location of the pathogenicity-modulating domain of citrus exocortis viroid. *EMBO J.* **5**, 2051-2055.

Zuker, M. 1989. On finding all suboptimal foldings of an RNA molecule. *Science* **244**, 48-52.

ABBREVIATIONS: CEV, citrus exocortis viroid; CSV, chrysanthemum stunt viroid; PCR, polymerase chain reaction; PSTV, potato spindle tuber viroid; RNase, ribonuclease; TASV, tomato apical stunt viroid; and TPMV, tomato planta macho viroid.

DISCUSSION OF R. OWENS' PRESENTATION

P. Palukaitis: As regards the pathogenicity domain, there is a region in exocortis viroid that can be linked with induction of severe symptoms in tomato. However, a strain of exocortis viroid that produces a severe symptom in tomato doesn't necessarily produce a severe symptom in citrus. Can you comment on the effect of host on pathogenicity?

R. Owens: We really don't know anything about that. The tomato assay system is artificial and even when we understand more about pathogenicity in tomato we will still probably be a long way from understanding pathogenicity in other agriculturally important plants, such as citrus.

W. Gerlach: Others have suggested recently that viroids may interact with 7S RNA and that this may have some role in symptom development. Can you comment on that? Secondly, is there a protoplast infection system where you could study viroid replication without the need for cell-to-cell movement, and use this to analyse the viroid mutants?

R. Owens: In answer to the second question, you can study replication in protoplasts with some difficulty and with quite slow kinetics. We have done experiments along those lines. The mutant with the change in the right terminal loop seemed to have much reduced infectivity in these experiments compared to the wild type. As regards the possible interaction with 7S RNA, none of the mutations I described would affect the base pairing between 7S RNA and the viroid. The mutations are adjacent to the region that would base pair, so if that model is correct it would have to operate indirectly, by the mutations making that region less stable or more stable, and so increasing or decreasing its availability to base pair. Going back to the question of differences in severity of symptoms induced, you would have to postulate that the 7S RNA in closely related species or cultivars would be significantly different. I would be surprised if that were so. It has got to be more complicated than that.

P. Palukaitis: Although the severity of symptoms of the same strain of PSTV can differ in different tomato cultivars, it seems unlikely that there is much difference between the 7S RNA species in these plants, particularly as H. Sanger has reported that several taxonomically diverse species, including *Gynura aurantiaca*, *Cineraria sp.* and tomato, have 7S RNAs which are virtually identical in nucleotide sequence. In addition, chrysanthemum stunt viroid, which produces no symptoms in tomato, binds to 7S RNA like PSTV, which causes obvious disease. This does not support the idea that binding to 7S RNA is an important factor in determining symptom severity, and so I agree with your comment.

B. Harrison: Is there any evidence of single gene differences in the plant host that affect reaction to viroids? Such systems might give a lead to what the viroid interacts with.

R. Owens: As far as I am aware, none. One problem is that there are few examples of a viroid causing a suitable spectrum of symptoms in different genotypes of the same plant species, for example, necrotic local lesions versus a systemic reaction. So there is little basis for studies of the host genetics of pathogenicity.

A. Siegel: In your random mutagenesis tests you only got mutants that were less, or equally, severe when compared with the wild type viroid. Were you surprised not to get any that were more severe? Nitrous acid would change C to U and A to G. I wonder if you used any other agent, such as hydroxylamine, which would cause other kinds of nucleotide change.

R. Owens: Yes we did but we concentrated on the nitrous acid mutants. I would not want you to think that everything made them less severe. I have only given you data for two of the 19 infectious mutants that we are characterising because they fit well with Rose Hammond's oligonucleotide-directed mutagenesis in the same region. In a few months we should have determined the sequences and symptoms of some of the other mutants.

Use of Plant Virus Satellite RNA Sequences to Control Gene Expression

W.L. Gerlach, J.P. Haseloff, M.J. Young and G. Bruening

CSIRO Division of Plant Industry, GPO Box 1600, Canberra ACT 2601, Australia

Satellite RNAs of plant viruses are sometimes considered to be molecular parasites of the helper virus-host plant system. They are not required for virus infection and have little or no homology with either their helper virus or the host plant, but they are recognized by the virus such that they replicate to high levels and become encapsidated in virus coat protein.

The efficiency of their multiplication in the virus infection system is the result of a number of streamlined biochemical functions. The small satellite RNAs must contain signals for recognition by the encapsidation process, participation in the replication system and modification of symptom development in infected plants. It is through a molecular dissection of the satellite RNA that we hope to characterize the sequences and structures associated with these properties. The elucidation of the molecular properties may also lead to the adaptation of these signals for new roles in molecular biology.

This presentation will be concerned with the satellite RNA of tobacco ringspot virus, the use of a reverse genetics approach to obtain information on the structure-function aspects of the molecule, the capacity of the satellite RNA for rapid evolution *in vivo* and the adapation of its self cleavage properties to produce new RNA enzymes.

SATELLITE RNA OF TOBACCO RINGSPOT VIRUS

The satellite RNA of the budblight strain of tobacco ringspot virus (abbreviated STobRV) has been cloned and sequenced (Buzayan *et al.*, 1986a). It is 359 bases long and can be found as linear and circular forms in infected tissues. Only one linear strand is encapsidated. This is designated "plus" strand, while the complementary strand which is only found in the infected plant is designated "minus" strand. It is thought that the circular forms which can be detected in infected cells are templates for a rolling-circle type mechanism for RNA transcription, giving rise to concatameric forms of plus and minus strand RNA during replication (Kiefer *et al.*, 1982). The plus and minus strand concatamers undergo self-catalysed cleavage, both *in vitro* (Prody *et al.*, 1986; Buzayan *et al.*, 1986b) and *in vivo* (Gerlach *et al.*, 1987) to produce unit length 359 base forms.

The presence of satellite RNA during TobRV infection leads to a marked amelioration of symptom development in host plants (Schneider, 1977; Gerlach *et al.*, 1986, 1987). Since STobRV does not appear to encode any functional polypeptides, its properties regarding replication, encapsidation and modification of symptom development must reside solely within its small RNA sequence. We are undertaking a reverse genetics approach to characterize regions of the STobRV genome associated with its biological and molecular properties.

MUTANTS OF STobRV ARE ABLE TO EVOLVE RAPIDLY IN VIVO

A library of linker insertion mutants of STobRV cDNA have been produced (Haseloff and Gerlach, 1989). These have small linker oligonucleotides inserted at an extensive array of sites across the STobRV sequence, along with associated duplications or deletions of bases flanking the sites of mutation.

A number of these mutants have been expressed from a bacterial plasmid expression vector, and used in coinoculations of cowpea seedlings along with TobRV. Analysis of sequences obtained from the infected plants after either 4 days or 26 days reveals the rapidity with which mutated STobRV sequences can evolve. About 25% of mutants applied to the plants contain satellite RNA sequences which are replicated and transmitted with the infection in the plant. In all those which have been assayed, however, the level of satellite RNA is lower than for wild type RNA controls and symptom development is greater. Presumably the decreased replication reflects the lower adaptive fitness of the mutants in propagation with the virus. In many instances, the satellite RNAs which are recovered differ from the input mutant sequence. This can be seen by size analysis of the recovered sequences; they have sizes different from both the input mutant sequence as well as native STobRV. When a particular mutant is inoculated onto different plants it may yield different size forms from different plants, and even within one plant different forms can sometimes be detected. It is important to realize that this can happen very rapidly in plants, with these mutants being detectable as early as 4 days after inoculation, although this most likely represents many rounds in terms of cellular RNA multiplication and generation.

To further characterize the molecular events in the evolution of these mutant sequences, samples of RNAs were isolated and amplified for sequencing by the PCR technique. The samples were taken from plants inoculated with a particular mutant STobRV designated D-25, which had a linker insertion after base 273 of the STobRV sequence along with concomitant loss of bases 274-282 (Haseloff and Gerlach, 1989). Analysis of a number of recovered sequences showed that base changes had occurred at and around the mutation site. As well, some sequences had changes at locations distant from the original mutant site. We assume that these are compensatory mutations which interact to functionally compensate, at least in part, the original mutation. The continued evolution of these modified forms is being studied by further propagation through serial passaging in cowpeas. We will be particularly interested to determine the nature of stable forms as they are produced.

In these experiments the mutant forms must be capable of both replication and encapsidation to be recovered. In order to characterize the inability of the majority of the mutants to propagate in the infected plants, these sequences are being analysed for the various molecular properties of STobRV both *in vitro* and in transgenic plants. The use of transgenic plants provides an alternative system for production of plus and minus strands and intercellular spread. Different lines have been established, each of which expresses a range of plus and minus sequence mutants, and their analysis has commenced.

ADAPTATION OF THE STobRV AUTOCATALYTIC CLEAVAGE MECHANISMS TO PRODUCE NEW ENZYMATIC RNAs

As mentioned earlier, the plus and minus strand sequences of STobRV are found in plants as multimeric forms, most likely produced by rolling circle mechanims from circular templates. These autolytically cleave at precise locations to produce unit length monomers with characteristic 5'-OH and 3' cyclic phosphate ends. The sequences required for autolytic cleavage of the RNAs have been characterized and provide the basis for the production of ribozymes which can be designed to cleave target RNAs at selected target sites.

For the plus strand, the initial observation of autolytic cleavage (Prody *et al.*, 1986) together with subsequent analysis of cloned segments of the sequence, led to the characterization of a small oligonucleotide segment which was able to self-cleave (Buzayan *et al.*, 1986b). This reaction requires only the presence of a divalent cation such as Mg^{++} and a neutral or higher pH. Similar self cleavage reactions have been observed in a number of other plant virus satellite RNAs (Forster and Symons, 1987a,b), a viroid (Hutchins *et al.*, 1986) and a cellular RNA transcript in newt (Epstein and Gall, 1987). Certain conserved bases and a secondary structure motif, termed a "hammerhead" (Forster and Symons, 1987a), have been associated with the sequences required for cleavage for each of these. Although the hammerhead secondary structure provides a convenient form for representation of the secondary structure of these sequences, it is likely that the mechanism for cleavage involves the formation of a complex tertiary entity which allows reactive groups to be brought into close proximity to the susceptible bond so that cleavage is catalyzed.

The *in vitro* mutants provided the stimulus for the development of new catalytic RNAs with the capacity to cleave specific target sites in new substrate RNAs (Haseloff and Gerlach, 1988). One mutant, designated D-51, had its linker insertion within the domain required for self cleavage of plus strand multimers, but did not affect the self cleavage reaction *in vitro*. This led to a split of the normal monomolecular reaction into a particular bimolecular reaction, with most of the conserved sequences required for the reaction on the catalytic segment (termed "ribozyme") and only a requirement for a GUC trinucleotide in the target sequence (or more likely XUY, where X= any base, Y=A,U or C). This structure is shown in Fig. 1. These minimal requirements in the target sequence meant that new ribozymes could be synthesized with specificity against virtually any natural target RNA, simply by assigning the appropriate sequence arms (for binding to the target by specific base pairing) to be associated with the catalytic domain of the new ribozyme.

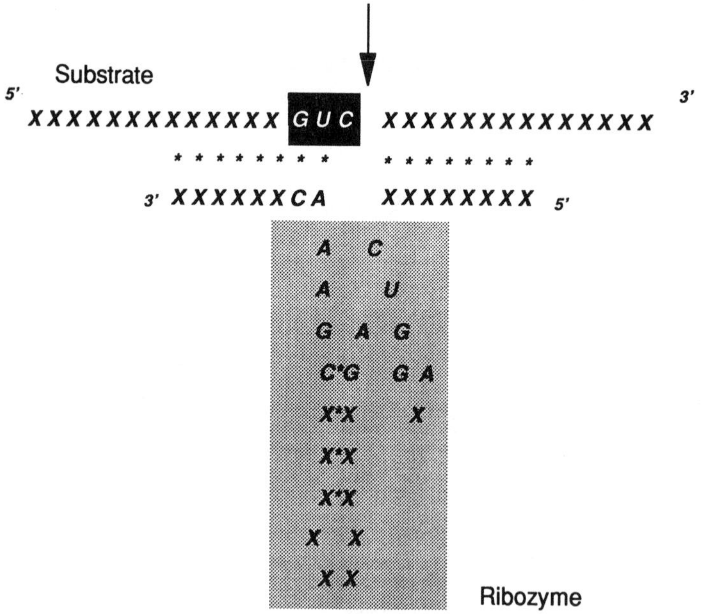

Fig. 1 Design of ribozymes based on the autolytic cleavage mechanism of plus strand of satellite RNA of tobacco ringspot virus. An RNA target sequence is chosen with a "GUC" motif, and "X" being any other base (s), as substrate. Cleavage (arrowed) occurs at the 3' side of the GUC as shown, when the substrate is reacted with the ribozyme. The ribozyme has two essential parts (1) the catalytic domain (shaded region) and (2) the arms which base pair with the target sequence in the substrate.

This concept was first tested *in vitro* against specific target sites within the mRNA for chloramphenicol acetyltransferase (CAT) (Haseloff and Gerlach 1988) but has subsequently been verified against a range of target RNAs including tobacco mosaic virus (TMV) RNA and citrus exocortis viroid RNA. The ability to cleave a viroid is significant, since it means that these ribozymes can be designed against even highly secondary structured molecules. The kinetics of reactions against such targets are being determined.

Whereas the plus strand cleavage occurs after the sequence GUC, the minus strand cleaves after ACA in the apparent absence of the conserved motif found in other RNAs described above. It has been demonstrated that the sequence requirements for the minus strand cleavage reside in two separate parts of the

STobRV sequence. There is a minimal sequence required around the cleavage site, with the cleavage being effected by a separate sequence domain in the STobRV minus strand (Haseloff and Gerlach, 1989; Feldstein *et al.*, 1989). Based on this, Hampel *et al.*, (1989) have adapted the minus strand cleavage mechanism to recognize new target RNA sequences. This is done in a similar way to that described for the ribozymes based on the plus strand, i.e., base sequence complementarity is used to target and deliver a catalytic ribozyme domain to a specified substrate RNA.

RIBOZYMES IN VITRO

The efficiency of ribozymes based on the plus strand STobRV have been studied *in vitro*. In particular, experiments have involved testing alterations in the three basic components of the system:

(a) Choice of target sequences. Our initial experiments involved selection of GUC trinucleotide targets in the substrate sequence since this is the sequence most often used at this location in the natural hammerhead sequences. However, GUA is used in one natural hammerhead, and we have subsequently found that both GUA and GUU but not GUG can be targeted in the CAT gene. This is in agreement with Koizumi *et al.*, (1988) who studied a natural hammerhead sequence and found that the sequence XUY (where X= any base, Y=A,U or C) can be adjacent to the cleavage site, so long as certain other compensatory changes are made for the altered "X" in XUC targets. Thus, it appears that the only sequence requirement of the substrate is a U as the second base upstream of the cleavage site and that target sites can be simply written XXXXXXXXXUYXXXXXXXXX where X= any base, Y=A,U or C.

(b) Catalytic Domain. The catalytic domain which we have routinely used has been that of STobRV. However, we have found that the domain used by the satellite RNA of subterranean clover mottle virus can also be used to produce ribozymes of similar activity. Furthermore, the extent of the base paired stem in the catalytic domain has also been altered, so that increase to 8 base pairs still maintains active ribozymes.

(c) Extent of base pairing. Initial observations were made using 8 base arms, so that complementarity was 16 bases (i.e., 2 x 8 bases). When tested at $37^{o}C$ this construction was approximately 20-fold less active than at $50^{o}C$. However, raising complementarity to 12 bases increased the reaction rate approximately 10-fold. Thus, the strategy for most of the subsequent experiments has been to increase complementarity even further, so that new constructs can be considered to be catalytic antisense molecules. These are long antisense sequence constructions which have multiple ribozyme domains inserted into them by mutagenesis. We consider that the complexity of the ribozyme sequences thus produced will provide for enhanced pairing of the ribozymes, while the multiple catalytic domains will cause extensive cleavage.

RIBOZYMES IN VIVO

In a series of experiments we have compared the *in vivo* effects of ribozymes on gene activity in both transient and transgenic studies.

A catalytic antisense construction with approximately 800 base complementarity has been used in transient assays in *Nicotiana tabacum* protoplasts. This construction targets the CAT gene. In experiments in which separate plasmids containing the CAT gene along with the ribozyme construction or comparable antisense control (each with the CaMV 35S promoter) are co-electroporated into protoplasts, the ribozyme has consistently had a greater effect on CAT gene expression. However, results from the experiments are too variable to allow pooling of data.

Experiments are underway with transgenic plants using the same gene constructions. The parent plant for these experiments was a transgenic which had the CAT gene originally introduced with the kanamycin resistance selection marker. In one experiment the ribozyme and comparable antisense controls were introduced using an *Agrobacterium rhizogenes* construction, selecting for transformed hairy roots and subsequently regenerating plants from these. Five plants were regenerated for each of the antisense gene and ribozyme gene insertions, as well as for five controls in which *Agrobacterium rhizogenes* transformation was done using the vector plasmid without insert. Results showed that high levels of CAT activity were maintained in the control plants, whereas the antisense and ribozyme constructions affected CAT gene activity. The distributions of CAT activity were overlapping, but the average for the ribozyme transformed plants was lower than for the antisense transformed plants, and there was one ribozyme transformant which had a 98% reduction in CAT activity.

Ribozyme gene constructions targetted against TMV have also been transformed into NN genotype *N. tabacum*. Comparable antisense controls were also included. We have tested multiple clonal plants from two independent transformants for each gene construction, as well as an untransformed control line (Fig. 2). The antisense plants did not show any apparent resistance to TMV compared to the untransformed control at high levels of inoculation. However, at these inoculation levels the plants transformed with the ribozyme constructions did show a degree of virus resistance, with the more active transformant having greater resistance. These experiments were done on clonal plants from primary regenerants after transformation. We have now collected seed from these and a range of other transformants which will be used for a more extensive analysis.

One aspect which we are only beginning to address is the question of delivery of ribozymes in transgenic plants. All of the experiments done so far have been done with gene constructions involving simply the ribozyme being driven by the CaMV 35S promoter. That this may be too simplistic is suggested by an experiment which has been performed in African Green Monkey COS

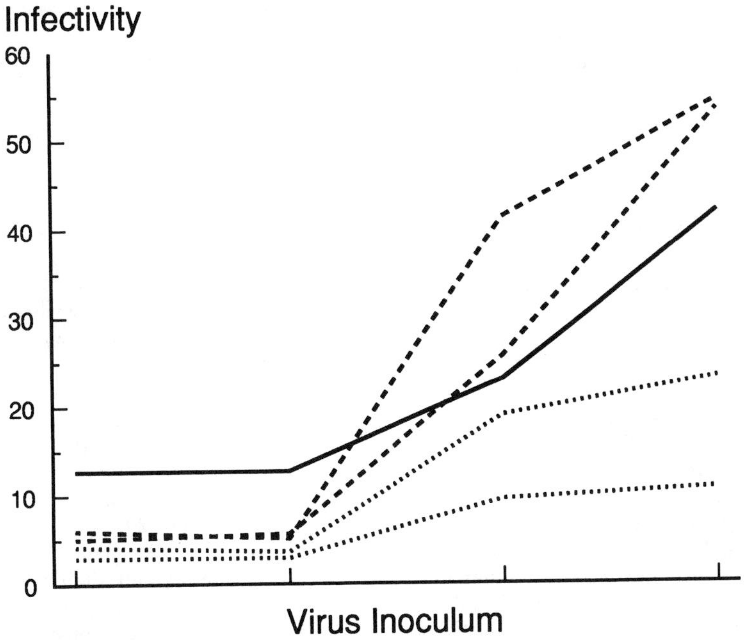

Fig. 2 Effect of ribozyme and antisense constructions on TMV infectivity in transgenic tobacco. Infectivity, expressed as relative numbers of lesions per leaf, is plotted against increasing virus inoculum for 5 different genotypes. Solid line: NN Samsun control; Dashed lines: two independent lines of NN Samsun transformed with tDNA to express antisense sequence against TMV; Dotted lines: two independent lines of NN Samsun transformed with tDNA to express ribozyme sequence against TMV.

cells (Jennings and Cameron, 1989). They used the ribozyme sequence which we had prepared targetted at the CAT2 site (Haseloff and Gerlach, 1988) with 2 x 12 base arms. Its activity was compared with the comparable 24 base antisense control sequence driven by the same adenovirus late promoter. When the sequences were directly transcribed from the input plasmid neither affected CAT activity in their transient assay experiments. However, when the ribozyme was part of a translated message sequence it did affect CAT activity significantly. The comparable antisense control did not affect activity. Therefore, we must now consider carefully future constructions which are to be prepared for plant transformation experiments.

CONCLUSIONS

We are working with a satellite RNA from a plant virus, tobacco ringspot virus. Although only 359 bases long, it has a number of properties associated with it including replication signals, encapsidation capability and a marked amelioration of symptom development in infected plants. A reverse genetics approach is being used to dissect these properties. From experiments so far, we have found that the satellite RNA has the ability to undergo rapid sequence changes under selection *in vivo*. Also, the autolytic cleavage activity has been adapted for the design of new ribozyme sequences. These have the potential for control of a range of targetted gene and viral functions *in vivo*.

REFERENCES

Buzayan, J.M., Gerlach, W.L., Bruening, G., Keese, P., and Gould, A.R. (1986a). Nucleotide sequence of satellite tobacco ringspot virus RNA and its relationship to multimeric forms. *Virology* **151**, 186-199.

Buzayan, J.M., Gerlach, W.L., and Bruening, G. (1986b). Satellite tobacco ringspot virus RNA: a subset of the RNA sequence is sufficient for autolytic processing. *Proc. Natl. Acad. Sci. USA* **83**, 8859-8862.

Epstein, L.M. and Gall, J.G. (1987). Self-cleaving transcripts of satellite DNA from the newt. *Cell* **48**, 535-543.

Forster, A.C. and Symons, R.H. (1987a). Self-cleavage of virusoid RNA is performed by the proposed 55- nucleotide active site. *Cell* **50**, 9-16.

Forster, A.C. and Symons, R.H. (1987b). Self-cleavage of plus and minus RNAs of a virusoid and a structural model for the active sites. *Cell* **49**, 211-220.

Gerlach, W.L., Buzayan, J.M., Schneider, I.R. and Bruening, G. (1986). Satellite tobacco ringspot virus RNA: biological activity of DNA clones and their in vitro transcripts. *Virology* **151**, 172-185.

Gerlach, W.L., Llewellyn, D. and Haseloff, J. (1987). Construction of a plant disease resistance gene from the satellite RNA of tobacco ringspot virus. *Nature* **328**, 802-805.

Hampel, A. and Tritz, R. (1989). RNA catalytic properties of the minimum (-)sTRSV sequence. *Biochemistry* **28**, 4929-4933.

Haseloff, J. and Gerlach, W.L. (1988). Simple RNA enzymes with new and highly specific endoribonuclease activities. *Nature* **334**, 585-591.

Haseloff, J. and Gerlach, W.L. (1989). Sequences required for self-catalysed cleavage of the satellite RNA of tobacco ringspot virus. *Gene* **82**, 43-52.

Hutchins, C.J., Rathjen, P.D., Forster, A.C. and Symons, R.H. (1986). Self-cleavage of plus and minus RNA transcripts of avocado sunblotch viroid. *Nucleic Acids Res.* **14**, 3627-3640.

Jennings P. and Cameron F. (1989). Specific gene suppression by engineered ribozymes in monkey cells. *Proc. Natl. Acad. Sci. USA*, 9139-9143.

Kiefer, M.C., Daubert, S.D., Schneider, I.R. and Bruening, G.B. (1982). Multimeric forms of satellite of tobacco ringspot virus RNA. *Virology* **121**, 262-273.

Koizumi, M., Iwai, S. and Ohtsuka, E. (1988). Construction of a series of self-cleaving RNA duplexes using synthetic 21-mers. *FEBS LETT.* **228**, 228-230.

Prody, G.A., Bakos, J.T., Busayan, J.M. Schneider, I.R. and Bruening G. (1986). Autolytic processing of dimeric plant virus satellite RNA. *Science* **231**, 1577-1580.

Schneider, I.R., (1977). Defective Plant Viruses. pp200-219 in Romberger, J.A. (Ed.) Beltsville Symposia on Agricultural Research - Virology in Agriculture (Allenheld Osmun, NJ).

DISCUSSION OF W. GERLACH'S PRESENTATION

B. Harrison: Which sequences in TMV RNA did you target in your experiments?

W. Gerlach: We used ribozymes designed to cleave the positive strand RNA of the U1 strain of TMV. Unfortunately I cannot give full details of the constructions that were used.

J. Hammond: With anti-viral ribozymes, does it make any difference whether you target the messenger-sense or the anti-sense RNA and, where you have long anti-viral constructs which are, in effect, anti-sense molecules with added ribozyme sites, do you know whether the cleaved sequence persists in cells as double-stranded RNA?

W. Gerlach: I have described ribozymes designed to target messenger-sense RNA. We have no results yet on ribozymes which target the negative strand. Nor do we know what is happening within the plant. Experiments with inoculated protoplasts may be informative and we hope to do these.

P. Palukaitis: What is the effect of mutating the first residue at the inner end of the arm which is bound to the ribozyme? Would a mutation in that residue prevent cleavage occurring?

W. Gerlach: We are now doing these tests with the CAT gene, and changing each base to every other base, alone and in combination. It appears that the G of the GUC target can be changed as long as a compensating change is made in the ribozyme. However, there are conflicting data in the literature as to whether pairing of this base is necessary. The rationale for using multiple ribozymes against viruses is that if a mutation occurs at one site, cleavage should still occur at the other sites.

T. Hohn: Can you comment on the prospects for use of ribozymes in mammalian virology, or anti-virology?

W. Gerlach: There is much interest, as indicated by the fact that the NIH have called for a major initiative on the use of ribozymes for HIV. We will just have to wait and see what happens.

R. Hull: With the Bam-linker mutants of the satellite RNA, were you looking at the progeny that was encapsidated or were you looking at unencapsidated satellite RNA?

W. Gerlach: I suspect that we were looking at encapsidated satellite RNA because we examined material from systemically infected leaves, but we are not entirely sure. Further experiments to clarify this point are in progress with transgenic plants.

M. Zaitlin: In the TMV experiment, you presumably transformed NN tobacco plants and then challenged them. You would therefore be testing the effect on establishment of infection centers and on replication at these centers. Was there any effect on lesion size when plants were transformed with antisense constructs or ribozymes?

W. Gerlach: No obvious effect on lesion size was found.

M Zaitlin : Most RNA plant viruses do not have any nuclear association during the replication. Do you know where in the cell the ribozyme occurs in an active form?

W. Gerlach: No, we do not. The construction that was used contained the 35S promoter, the ribozyme sequence and a polyadenylation signal, so presumably the ribozyme gets out of the nucleus and into the cytoplasm and is available in both places. But better kinds of construction can probably be made.

C. Collmer: Does the satellite act, as a ribozyme, on the RNA of the helper virus?

W. Gerlach: I am not aware of anyone looking but I suspect it does not, because it normally works to cleave another part of the same molecule. When RNA of a satellite-containing virus isolate is isolated from virus particles there is no evidence that the genomic RNA species are cleaved, despite the preparation containing a mixture of genomic and satellite RNA.

P. Berger: What was the size of the arms in the TMV ribozyme?

W. Gerlach: The ribozyme was a catalytic anti-sense sequence of about a kilobase with multiple ribozyme domains.

O. Barnett: In some of the early work on trying to find the replicative complex of TRSV, short strands of double-stranded RNA were found. Is it possible that the satellite RNA cleaves double-stranded viral genomic RNA?

W. Gerlach: We have not done any tests but I doubt if it would act in that way.

R. Pacha: Have you made protoplasts from transgenic plants containing the catalytic TMV antisense construction and then infected them with TMV and looked at the level of replication?

W. Gerlach: That experiment is being done now, but I do not have the answer yet.

A. Miller: As the constraints on ribozyme design seem to be small - just a CUGA and a GA and then a few other bases - have you done database searches to see whether there are other naturally occurring RNA molecules which might by chance be able to act as ribozymes and to find out how often that motif occurs?

W. Gerlach: We have not done that. You could look for those conserved bases and the capacity to form a stem loop between them.

Infectious Synthetic Transcripts of Beet Necrotic Yellow Vein Virus RNAs and Their Use in Investigating Structure-Function Relations

Isabelle Jupin, L. Quillet, Ursula Niesbach-Klösgen, S. Bouzoubaa, K. Richards, H. Guilley and G. Jonard

Institut de Biologie Moléculaires des Plantes,
12 rue du Général Zimmer, 67084 Strasbourg Cédex, France

Beet necrotic yellow vein virus (BNYVV) is a soil-borne plus-strand RNA virus possessing a quadripartite genome. In the field BNYVV is transmitted in a persistent fashion by the plasmodiophoroid fungus *Polymyxa betae* and is responsible for a severe disease of sugar beet known as rhizomania, characterized by extensive proliferation of lateral roots, necrosis of vascular tissues and severe stunting of infected plants (Tamada, 1975; Brunt and Richards, 1989). The virus generally remains confined to the roots of naturally infected sugar beet but can be transmitted by mechanical inoculation of an infected root extract to leaves of *Chenopodium quinoa* or *Tetragonia expansa*, which are commonly used for propagation purposes.

GENOME ORGANIZATION

BNYVV virions are rigid ribonucleoprotein rods of helicoidal symmetry (Steven *et al.*, 1981) composed of multiple copies of the 21 kilodalton (Kd) coat protein and a single molecule of RNA. The rods are of constant diameter (20 nm) but fall into four discrete length classes, reflecting the length of the encapsidated RNA. The four genomic BNYVV RNAs are capped at their 5' termini and 3' polyadenlyated (Putz *et al.*, 1983). Regions of sequence homology among the viral RNAs are limited to their extremities (see below). Fig. 1 presents the disposition of long open reading frames (ORFs) on the genome RNAs. The 6.8 kilobase (Kb) RNA 1 contains a single long ORF encoding a 237 Kd polypeptide sharing domains of sequence homology with large nonstructural polypeptides implicated in the replication of other plus-strand RNA viruses of the Sindbis supergroup (for review, see Goldbach, 1987). The first domain (Fig. 1) contains a purine nucleoside triphosphate binding fold consensus sequence embedded in a longer sequence characteristic of proteins possessing DNA or RNA helicase activity (Gorbalenya *et al.*, 1988; Hodgman, 1988). The T/SXXXTXXXNT/S and GDD motifs found in the putative replicases of all plant and animal plus-strand RNA viruses (Kamer and Argos, 1984; Goldbach, 1987) are situated in the second homologous domain.

The cistron for the viral coat protein is located nearest the 5' extremity of the 4.6 Kb RNA 2 molecule (Fig. 1). The cistron terminates with a single UAG codon which undergoes suppression about one out of every ten times it is encountered by a translating ribosome to produce a 75 Kd readthrough protein with the coat protein sequence at its N-terminus (Ziegler *et al.*, 1985; Bouzoubaa *et al.*, 1986; Quillet *et al.*, 1989). Four additional ORFs occupy the 3' terminal half of RNA 2 (Fig. 1). The C-terminal portion of P42 (ORF 3), P13 (ORF 4) and to a lesser extent P15 (ORF 5) display sequence homology with succesive ORFs on RNA β of barley stripe mosaic virus (Gustafson and Armour, 1985; Bouzoubaa *et al.*, 1987) and the genomic RNA of potexviruses (Morozov *et al.*, 1987; Forster *et al.*, 1987). These similarities indicate that BNYV, barley stripe mosaic virus and the potexviruses are more closely related than hitherto suspected. Furthermore, the manner in which the homologous ORFs appear to have cosegregated in the course of evolution may betoken a functional linkage between the polypeptides. Finally, with regard to function it is noteworthy that the C-terminal region of P42 (as well as the homologous domains of the other viruses) contains a second purine nucleoside binding fold and associated helicase consensus sequences (Gorbalenya *et al.*, 1988; Hodgman, 1988).

BNYVV RNA 3 (1.8 Kb) and RNA 4 (1.5 Kb) possess ORFs for polypeptides of 25 Kd and 31 Kd, respectively (Fig. 1). In both cases the ORF is preceded by a long 5' noncoding region, 444 nucleotides (nt.) for RNA 3 and 379 nt. for RNA 4. These upstream regions contain additional short ORFs, unlike the situation with RNAs 1 and 2 in which the first AUG encountered marks the beginning of a long translationally active ORF. No sequence homology of note was detected beween the 25 Kd and 31 Kd polypeptides and proteins catalogued in Release 20 of the NBRF database.

BNYVV RNAs 3 and 4 are invariably present along with RNA 1 and 2 in naturally infected sugar beet roots but one or both small RNAs may suffer internal deletion or disappear entirely when isolates are maintained by mechanical inoculation to leaves (Koenig *et al.*, 1986; Kuszala *et al.*, 1986; Burgermeister *et al.*, 1986; Lemaire *et al.*, 1988). These observations indicate that RNAs 3 and 4 are not essential for the viral replication cycle *per se*, at least in leaves, but are indispensable for the natural infection cycle in soil, i.e., *P. betae*-mediated infection of beet root and proliferation within this tissue. Working out the role(s) of RNAs 3 and 4 in this process will be of considerable interest.

No counterparts of RNA 3 and 4 have been observed in other plant viruses. In particular, no comparable entities exist for the bipartite soil-borne wheat mosaic virus and peanut clump virus, well characterized members of the recently created furovirus (fungus-transmitted rod-shaped) group (Shirako and Brakke, 1984) with which BNYVV has been tentatively associated (Brunt and Richards, 1989).

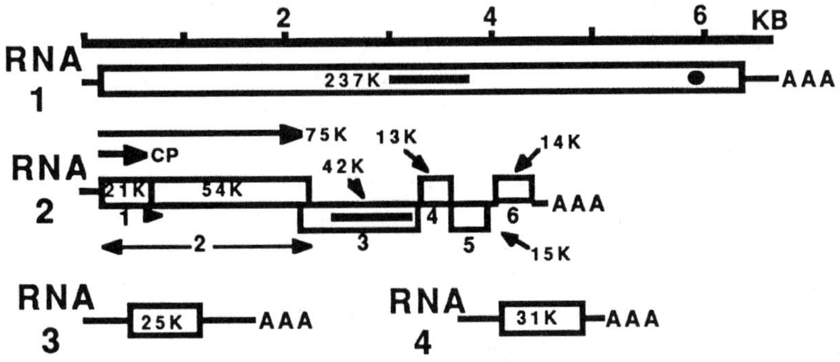

Fig. 1. Genetic organization of BNYVV RNAs (Bouzoubaa *et al.*, 1987). Long ORFs are represented by hollow rectangles with the molecular weight in kilodaltons of the largest possible translation product predicted for each ORF indicated within or nearby. Arrows above RNA 2 represent translation of the coat protein (CP) and the 75 Kd readthrough protein produced by suppression of the termination codon indicated by the triangle. Heavy bars represent regions of amino acid sequence homology with helicases and the black circle the "GDD" replicase consensus sequence (see text).

EXPRESSION AND SUBCELLULAR LOCATION OF BNYVV PROTEINS

In order to follow the appearance of nonstructural BNYVV proteins in infected tissue we have raised polyclonal rabbit antisera against synthetic peptides or translational fusions corresponding to all of the long ORFs predicted from the sequence of RNAs 2, 3 and 4. Synthetic oligopeptides were used to obtain antisera specific for the readthrough portion of the 75 Kd polypeptide and P13 of RNA 2. Portions of the other ORFs were inserted as in-frame fusions behind the N-terminal part of cloned bacteriophage lambda CI protein and purified from extracts of transformed *Escherichia coli* (John *et al.*, 1985). Using these antisera in Western blot experiments, polypeptides corresponding to all of the viral ORFs except P15 (ORF 5) of RNA 2 have been detected in extracts of infected *C. quinoa* leaves (U. Niesbach-Klösgen, unpublished observations). The viral polypeptides were first detected 5-6 days post-infection, the time when local lesions first appear, and remained present until at least 11 days post-infection. It is of particular interest that both the RNA 3 and RNA 4 polypeptides, P25 and P31, are expressed in view of their long 5' noncoding regions and the fact that neither RNA is essential for virus infection of leaves.

To investigate the subcellular location of the different virus-coded polypeptides the infected leaf homogenates were fractionated (Godefroy-Colburn *et al.*, 1986) into a 1000g pellet (nuclei and chloroplasts), 30,000g pellet (membranes), 30,000g supernatant (cytosol) and a cell wall-enriched fraction consisting of particulate material retained on a nylon screen after three washes with a buffer containing 2% Triton X100. Coat protein and the 75 Kd readthrough protein were present in all subcellular fractions but were usually most abundant in the 30,000g pellet. P42 and P13 were found primarily in the 30,000g (membrane) fraction although small amounts were also present in the cell wall fraction. The other viral-coded polypeptides were located exclusively in the cytosol. Thus, with the possible exception of the as of yet undetected P15 polypeptide of RNA 2, none of the BNYVV polypeptides studied so far are tightly associated with the cell wall fraction in a manner comparable to the viral proteins thought to be involved in cell-to-cell movement of viruses such as alfalfa mosaic virus (Godefroy-Colburn *et al.*, 1986; Stussi-Garaud *et al.*, 1987), tobacco mosaic virus (Tomenius *et al.*, 1987; Moser *et al.*, 1988) and cauliflower mosaic virus (Albrecht *et al.*, 1988; Linstead *et al.*, 1988).

INFECTIOUS TRANSCRIPTS

Cell-free transcription systems using a purified RNA polymerase and viral cDNA positioned behind a cloned bacteriophage promoter sequence have been used successfully to generate infectious run-off transcripts for a growing number of plant viral RNAs (Ahlquist *et al.*, 1984; Dawson *et al.*, 1986; Meshi *et al.*, 1986; Allison *et al.*, 1988; Vos *et al.*, 1988; Domier *et al.*, 1989; Hamilton and Baulcombe, 1989; Heaton *et al.*, 1989; Petty *et al.*, 1989; Weiland and Dreher, 1989). Preliminary studies with transcripts of RNA 3 and 4 indicated that short nonviral sequences following the 3' poly(A) tail have relatively little effect on transcript infectivity (Ziegler-Graff *et al.*, 1988). Extra nucleotides at the 5' terminus, on the other hand, were more harmful, as has proven to be the case for other transcribed viral RNAs (Dawson *et al.*, 1986; Ahlquist *et al.*, 1987). Consequently, the BNYVV cDNA plasmids for transcription were engineered to contain only one or two extra G residues preceding the viral RNA sequence proper. Appropriately modified transcription vectors containing full-length cDNAs of RNAs 1, 3 and 4 of our type isolate F3 were obtained without any particular difficulty (Ziegler-Graff *et al.*, 1988; Quillet *et al.*, 1989) but construction of a plasmid containing full-length cDNA of RNA 2 proved problematic. Two overlapping approximately half-length cDNA clones harboring the 5' and 3' terminal regions were available from earlier work (Bouzoubaa *et al.*, 1986). The region of sequence overlap of the two inserts contained a unique BstXI site. After introduction of a T7 promoter sequence upstream of the viral cDNA 5' terminus the two fragments were fused *via* this site. This step took place efficiently, as judged by agarose gel electrophoresis of the ligation products, but all attempts to amplify plasmids containing the joined fragments met with failure (Quillet *et al.*, 1989) due, apparently, to toxicity of the full-length cDNA. (This in spite of the fact that neither of the shorter component fragments was toxic to bacteria when present alone in plasmids!) Possibly, one of the cDNA fragments encodes a polypeptide whose expression is lethal to bacteria while the other contains a sequence which, by chance, has procaryotic promoter activity.

The aforesaid circumstances compelled us to carry out transcription directly on the ligated cDNA rather than first attempting to amplify a plasmid containing the ligation product in bacteria. The 5' and 3' extremities of the fragments to be joined were treated with phosphatase in the course of their preparation in such a way as to minimize unwanted ligations but without removing the terminal phosphates from the BstXI sites. Note that BstXI gives rise to nonpalindromic sticky ends so that ligation at this site can occur only in the appropriate orientation. The desired linear full-length cDNA was produced in high yield and proved to be an efficient template for *in vitro* transcription, giving rise to a transcript with the mobility expected of full-length RNA 2 (Quillet *et al.*, 1989).

BIOLOGICAL PROPERTIES OF THE TRANSCRIPTS

When a mixture of the four capped BNYVV RNA transcripts was inoculated to leaves of *C. quinoa* or *T. expansa* bright yellow local lesions similar to the symptoms of infection with natural BNYVV appeared 6-8 days post-infection (Quillet *et al.*, 1989). The specific infectivity of the transcript mixture, while sufficient for our purposes, was 10-50 times lower than authentic BNYVV RNA, presumably due to incomplete capping of the transcripts and/or inhibitory effects of the terminal nonviral sequences. Progeny virus and RNA obtained from transcript-infected tissue had infectivity similar to that of natural virus and RNA and the 5' terminal nonviral G or GG residues present in the original transcripts were absent. The fate of the nonviral 3' sequence has not been investigated.

The BNYVV isolate derived from the four synthetic transcripts will be referred to as Stras 1234. Neither the RNA 1 nor the RNA 2 transcript was infectious to leaves when inoculated alone. Mixtures containing only RNA 1 plus RNA 2 (Stras 12) or these two RNAs in combination with one or the other of the small RNAs (Stras 123 and Stras 124) were also tested by inoculating the appropriate combination of transcripts and were found to be infectious. The infectivity of these combinations affords formal proof that not even trace amounts of the small RNAs are necessary for BNYVV multiplication on leaves. Such transcript-derived isolates will be extremely useful not only because of their rigorously defined RNA composition but also because they provide a constant RNA 1 plus 2 background for study of the influence of the small RNAs on symptom expression and other properties.

As observed previously with natural BNYVV isolates (Kuszala *et al.*, 1986; Koenig and Burgermeister, 1989) the presence of RNA 3 in the transcript-derived isolates has pronounced effects on symptom expression, producing bright yellow local lesions (yellow spot (YS) symptoms; Tamada, 1975) on inoculated leaves of *C. quinoa* and *T. expansa* (Table 1). RNA 4 has less pronounced effects. The isolates Stras 12 and Stras 124 which lack RNA 3 give rise to faint concentric rings or pale green spots (Table 1). On *T. expansa* the symptoms produced by these latter two isolates are often virtually imperceptible in natural light although lesions are readily visible on inoculated leaves viewed under ultraviolet illumination (I. Jupin, unpublished observations).

Table 1. Symptoms induced by BNYVV isolates of different RNA composition obtained from synthetic transcripts.

Isolate	RNA composition	Symptoms† *C. quinoa*	*T. expansa*	Spinach
Stras 1234	RNA 1 + 2 + 3 + 4	YS*	YS	systemic
Stras 123	RNA 1+ 2 + 3	YS	YS	systemic
Stras 124	RNA 1+ 2 + 4	CS*	CS	systemic≠
Stras 12	RNA 1 + 2	CR or CS	CR	systemic≠

† infection of *C. quinoa* and *T. expansa* confined to inoculated leaves ; YS : yellow spot (intense yellow lesions) ; CS : chlorotic spot (pale green local lesions) ; CR : concentric rings.

* center of lesion sometimes became necrotic.

≠ no symptoms on upper noninoculated leaves.

Spinach (*Spinacia oleracea*) is a reliable systemic host for BNYVV in our hands (Lemaire *et al.*, 1988). Expanding yellow local lesions appear on inoculated leaves about 7 days post-infection and virus can be detected in the upper noninoculated leaves and in the roots 2-3 weeks later. The BNYVV isolates derived from the synthetic transcripts reproduced this behavior with systemic spread to the upper leaves and roots occuring regardless of the RNA 3 and 4 composition. Stras 1234, Stras 123 and natural BNYVV isolates containing RNA 3 gave rise to deformation and heavy mosaic on noninoculated leaves while the two Stras isolates from which RNA 3 was absent produced a symptomless infection on the upper leaves (Table 1).

A COAT PROTEIN MUTATION AFFECTING RNA PACKAGING

While screening candidate cDNA clones containing the 5' terminal portion of RNA 2 for use in construction of synthetic transcripts we fortuitously detected a clone with a single base substitution at nt. 499, resulting in the replacement of arginine by serine at position 119 of the coat protein (Quillet *et al.*, 1989). The clone pB25 containing this sequence variant was singled out because cell-free translation of its synthetic transcript produced coat protein of slightly lower electrophoretic mobility than that observed with natural BNYVV RNA 2 or the synthetic transcripts of other RNA 2 cDNA clones. There were no other sequence changes in the variant clone. Direct analysis of cDNA obtained from natural virion RNA using a 5' labelled synthetic oligodeoxynucleotide primer revealed no detectable sequence heterogeneity at nt. 499, suggesting that the base substitution is either a cloning artifact or that the variant sequence is present in only a small fraction of the natural RNA population.

A full-length RNA 2 transcript based on pB25 was infectious when inoculated to leaves of *C. quinoa* in combination with transcripts of RNAs 1, 3 and 4 (Quillet *et al.*, 1989). The resulting local lesions were similar in appearance to those obtained with wild-type RNA 2 transcript. Viral coat protein with the characteristic anomalous electrophoretic behavior was easily detected on Western blots of protein extracted from the lesions but no virus particles were visible in leaf dips. A crude homogenate prepared from the lesions was noninfectious and contained no viral RNA detectable by dot-blot using a mixture of probes specific for all four RNAs. A phenol extract, on the other hand, contained the full complement of intact viral RNAs and was infectious. We conclude that infection occurs and that viral coat protein is synthesized and can accumulate in leaves inoculated with the pB25 sequence variant but that there is a defect in RNA encapsidation so that the viral RNAs are degraded when exposed to nucleases present in plant sap. It is noteworthy that RNAs 1, 3 and 4 as well as the variant RNA 2 are degraded in the crude plant extract, ruling out the possibility that the point mutation on RNA 2 has hit a packaging signal on this RNA rather than affecting the properties of the coat protein itself.

When a transcript mix containing the pB25 variant RNA 2 sequence was inoculated to spinach, lesions appeared on the inoculated leaves but no viral RNA was detected in phenol extracts of noninoculated upper leaves or roots 4 weeks post-infection (Quillet *et al.*, 1989). Thus it appears that efficient long distance movement of BNYVV, in spinach at least, requires a properly functioning viral coat protein.

INTERNAL DELETIONS IN BNYVV RNAS 3 AND 4

As noted above, BNYVV is customarily propagated on leaves of *C. quinoa* and many of the available laboratary isolates have been maintained for long periods of time in this host. It is often observed that such isolates accumulate deleted versions of RNA 3 and 4 or lose one or both small RNAs entirely (Kuszala *et al.*, 1986; Burgermeister *et al.*, 1986). The fact that such deletions, when they occur, give rise to a limited number of discrete species rather than a heterogenous population of shortened forms suggests that there are preferred sites ("hot spots") for deletion. All the deletions mapped by sequence analysis to date (Bouzoubaa *et al.*, 1985, 1988 and additional observations) lie within the interior of RNA 3 or 4, truncating or eliminating the long open reading frame (Fig. 2). Deletions extending too close to the extremities of the molecules are undoubtedly selected against because these regions harbor signals for replication and encapsidation (see below) whose loss would result in the disappearance of the deleted forms from the isolate.

Working with naturally occuring BNYVV isolates, it is difficult to assess how frequently deletions occur in a virus stock. Generation of a deletion could, for example, be a quite rare event but the shortened forms could persist in trace amounts in the isolate for long periods of time before being detected. Furthermore, except for RNA 3 of isolate F15, an isolate which derives from the type isolate F3, the sequences are not known for the parent forms of RNA 3 and 4 of the other isolates. Comparison of the sequences of the deleted RNAs 3 and 4 of isolates G1, G4 and F5 to the corresponding regions of RNA 3 and 4 of the type isolate reveals about 2 % interisolate variation (Bouzoubaa *et al.*, 1985, and additional observations). Thus the portions of RNA 3 and 4 which were eliminated in the aforesaid isolates almost certainly differed slightly from the type RNA 3 and 4 sequences with consequent uncertainties in modelling the details of the deletion phenomenon. Thus the synthetic transcripts offer a double advantage in studying the deletion process in that they represent a perfectly defined starting material while providing a means of "setting the clock to zero" with respect to the kinetics of deletion.

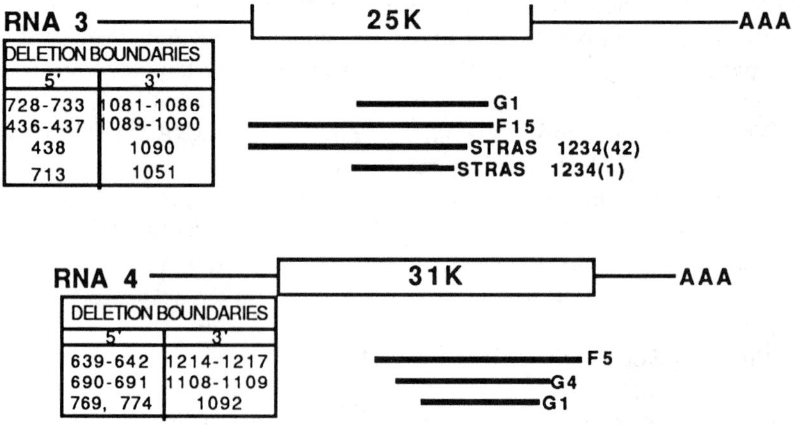

Fig. 2 Naturally occurring deletions within RNA 3 and 4 from different BNYVV isolates. The extent of each deletion is indicated by a heavy line and the boundary coordinates with respect to the isolate F3 sequence are given to the left. For many isolates, the exact limits of the deletion cannot be determined unambiguously because of sequence changes at the borders or terminal sequence redundancy. In these cases the possible range for each border is given.

After only one passage of Stras 1234 on *C. quinoa*, two specific deleted forms of RNA 3 and a deleted RNA 4 species have been detected (authors' unpublished observations). The shortened forms have electrophoretic mobilities similar to the major deleted species of RNA 3 and 4 which have accumulated in isolate F3 during multiple *C. quinoa* passages. Two deleted forms of RNA 3 present in the Stras isolate were characterized by sequence analysis of cDNA clones (authors' unpublished observations). As expected, the deletion boundaries fall within the interior of the RNA 3 sequence in both cases (Fig. 2). In the case of Stras 1234(42) the 3' border is identical to and the 5' border differs by only one nt. from the deletion boundaries observed for RNA 3 (F15).

The deletions in BNYVV RNAs 3 and 4 are most probably generated by a copy choice mechanism during RNA replication in which the viral RNA polymerase plus the nascent daughter strand switch from one point on the template to another. Such a mechanism is thought to account for the origin of defective interfering particles in negative strand RNA viruses (Lazzarini *et al.*, 1981) and for homologous recombination in picornaviruses (Kirkegaard and Baltimore, 1986). Models involving guide sequences (Kuge *et al.*, 1986), secondary structure (Romanova *et al.*, 1986) and zones of sequence homology (King, 1988) have been advanced to account for the locations of the cross-over sites. Although deletion hot spots in RNA 3 and 4 certainly exist, not enough examples have been mapped yet to permit critical features to be identified. No obvious consensus sequence is present at the deletion boundaries in the cases examined so far nor, with the exception of RNA 4 of isolate G1 (Bouzoubaa *et al.*, 1985), is there extended sequence homology near the deletion boundaries which might be expected to facilitate translocation of the nascent daughter strand. One interesting feature, however, warrants mention. A computer algorithm (Zuker and Stiegler, 1981) predicts that the central portion of RNA 3 (F3) will fold into a large branched hairpin structure. There is no evidence that such a structure exists in solution but it is none the less intriguing that the left- and right-hand boundaries of the deletions observed in RNA 3 (F3) are brought into close juxtaposition in this hairpin (authors' unpublished observations).

A SUBGENOMIC RNA DERIVED FROM BNYVV RNA 3

In addition to the deleted forms of RNA 3 and 4 described above BNYVV infection also gives rise to a species of about 600 nt. which appears to be a subgenomic RNA derived from RNA 3. An apparently identical small RNA 3-related species has also been observed by Burgermeister *et al.* (1986). The species in question, which we will refer to as RNA 3Sub, has the following properties (authors' unpublished observations): (1) it is detected by RNA 3-specific probes on Northern blots of total and polyadenylated RNA extracted from BNYVV-infected *C. quinoa* leaves and sugar beet roots but it is not encapsidated; (2) it is produced for all BNYVV isolates examined (including those derived from synthetic transcripts) provided that they contain RNA 3 but is absent if RNA 3 is absent; (3) hybridization with specific probes, RNase protection experiments and sequence analysis show that RNA 3Sub is colinear with the 3' terminal portion of RNA 3 with its 5' terminus located at about nt. 1231 (Fig. 3). Thus RNA 3Sub can be distinguished from the internally deleted forms of RNA 3 described above by the fact that it invariably appears upon infection whenever RNA 3 is present in the inoculum, that it is not encapsidated and that it does not contain the 5' terminal portion of the RNA 3 sequence. It will be shown below that an extended 5' terminal sequence is necessary for RNA 3 accumulation. Consequently, this last feature represents strong evidence that RNA 3Sub cannot exist autonomously but is synthesized *de novo* from RNA 3 in the course of infection. A small ORF encoding a 4.6 Kd polypeptide lies 45 nt. downstream of the RNA 3Sub 5' terminus (Fig. 3). It remains to be established if this or another polypeptide is expressed from RNA 3Sub and, if so, what role it might play in the infection process.

MAPPING REPLICATION SIGNALS ON RNA 3

Because BNYVV RNA 3 and 4 are not essential for RNA 1 and 2 replication in leaves, the small RNAs represent a useful material for mapping sequence domains involved *in cis* in their own replication. RNA 3 was chosen for the first such experiments because it contains numerous convenient restriction sites. A set of Bal 31 deletions starting with a unique centrally located AccI site and extending in steps toward the 3' terminus was engineered into the RNA 3 sequence at the DNA level and the resulting transcripts (Fig. 3) were inoculated to *C. quinoa* leaves along with BNYVV RNAs 1 and 2. The deleted transcripts multiplied efficiently in the infected leaves for all deletions extending up to and including nt. 1704 (variant B4ΔAS). Deletions eliminating 21 nt. or more beyond this point (variants B10ΔAS, B8ΔAS and B12ΔAS) reduced accumulation of RNA 3 to below the limits of detection. All of the deleted versions of RNA 3 which replicate are also encapsidated, as judged by their resistance to degradation in crude cell sap. Thus the sequences which initiate or otherwise control RNA 3 packaging are most probably located on the 5' distal portion of RNA 3, consistent with the fact that the 3' terminal subgenomic fragment, RNA 3Sub, is not encapsidated.

Extensive but not perfect sequence homology exists involving the 3' terminal 69 residues preceding the poly(A) tails of BNYVV RNAs 1-4 (Domain A in Fig. 4). Additional zones of sequence homology are found just upstream of Domain A between RNAs 1 and 2 (Domains D, E and F; Fig. 4) and between RNAs 3 and 4 (Domains B and C; Fig. 4). Domain A of all four RNAs can be folded into a pair of contiguous hairpin loops whose existence is supported by covariation of base changes in the base-paired stem regions of the hairpins (Jupin *et al.*, to be published). These conserved hairpin structures represent a plausible recognition site for the viral replicase involved in minus-strand RNA synthesis. The properties of the RNA 3 deletion variants tested are evidently consistent with such a model in that deletions which encroach upon Domain A such as B10ΔAS and B8ΔAS inhibit productive RNA 3 replication while deletions removing adjacent sequences (e.g. B4ΔAS) are without effect. In particular Domains B and C (Fig. 4) are unnecessary for RNA 3 accumulation and it will be a matter of considerable interest to discover what role these regions of RNA 3/4 homology have to play in the infection process.

DEL. LIMITS		REP	ENCAP
1036-1472	1A ΔAS	+	+
1036-1534	B42ΔAS	+	+
1036-1597	B19 Δ AS	+	+
1036-1652	B14Δ AS	+	+
1036-1687	B6Δ AS	+	+
1036-1704	B4 Δ AS	+	+
1036-1724	B10Δ AS	-	-
1036-1730	B8Δ AS	-	-
1036-POLY(A)	B12 Δ AS	-	-
933-1033	24 Δ XA	+	+
755-1033	1 Δ BA	+	+
570-1033	16 Δ HA	+	+
382-1033	1 Δ EA	+	+
260-1033	1 ΔXA	-	-
181-1033	1 ΔHA	-	-
71-1033	1 Δ NA	-	-
314-376	1 Δ SE	+	+
265-309	1Δ XS	-	-
181-260	1 ΔHX	-	-
71-180	1 ΔNH	-	-
382-1472	1 ΔES	+	+

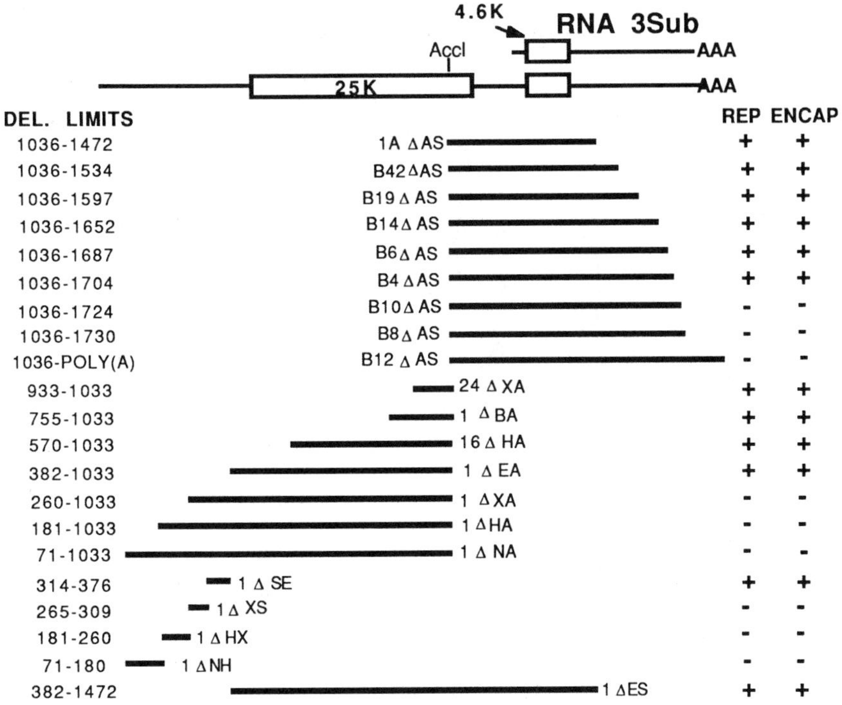

Fig. 3 Effect of deletions introduced into RNA 3 transcript on its accumulation and encapsidation in *C. quinoa* leaves. The extent of each deletion is indicated by a heavy line and the precise bounderies are given to the left. The location of the subgenomic species, RNA 3Sub, and of a small ORF which may be expressed from the subgenomic RNA are shown above the RNA 3 map.

In order to investigate the influence of the 3' poly(A) tail on BNYVV replication, an RNA 3 transcription vector was constructed from which the poly(A) tail was eliminated by synthetic oligodeoxynucleotide-directed mutagenesis (Jupin *et al.*, to be published). In the construct, the C residue at the 3' terminus of the heteropolymeric RNA 3 sequence was fused into a unique PstI site in such a way that PstI-linearized plasmid should produce an RNA 3 run-off transcript, t35A0, ending with this last C residue (Fig. 5). When inoculated to *C. quinoa* along with transcript-derived RNA 1 and 2, the biological activity of t35A0 was much lower than that of normal RNA 3. Most of the local lesions had the concentric ring or chlorotic spot phenotype characteristic of an RNA 1 plus 2 infection and no RNA 3 was detectable in these lesions. A few yellow spot lesions were also present, however, which proved to contain normal amounts of full-length RNA 3. About half of this RNA 3 bound tightly to an oligo-dT-cellulose column (natural RNA 3 behaves in an identical fashion) indicating that a substantial fraction of the progeny RNA 3 molecules have reacquired lengthy poly(A) tails. The nature of the

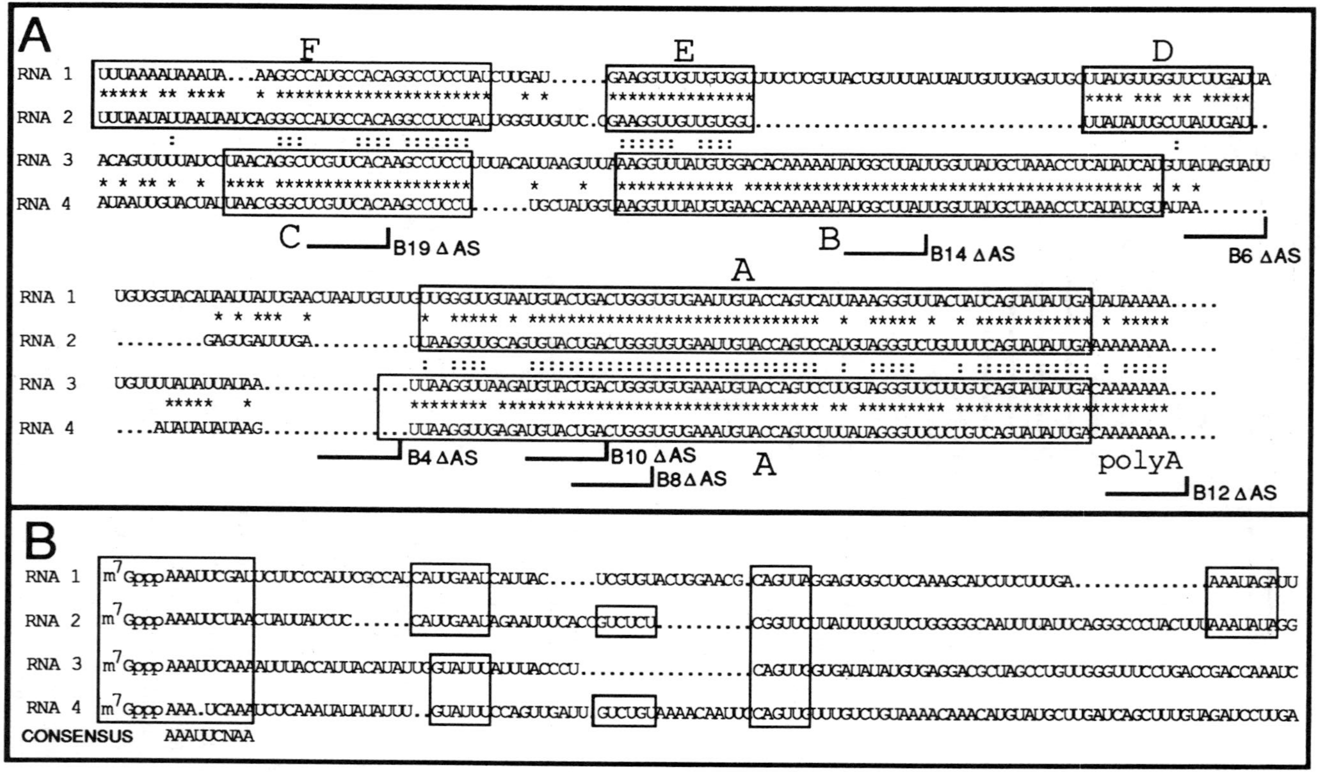

Fig. 4 Sequences near the 3' terminus (A) and 5' terminus (B) of BNYVV RNAs 1-4 (Bouzoubaa *et al.*, 1987). Gaps (represented by dots) have been introduced to maximize the alignments and regions of sequence homology are boxed. In (A) matches between RNAs 1 and 2 and between RNAs 3 and 4 are indicated by asterisks and matches betwen all 4 RNAs by colons. The right-hand boundaries of some of the RNA 3 deletions shown in Figure 3 are indicated beneath the aligned 3' terminal sequences. The 5' terminal sequence of RNA 3, previously reported to begin with 4 A residues (Bouzoubaa *et al.*, 1986), has been amended to A_3 (H. Guilley, unpublished observations) by analysis of dC- and dT-tailed reverse transcripts as described by Deborde *et al.*,(1986).

RNA 3 fraction which flows through the oligo-dT-cellulose column has not been investigated further but it probably consists of partial degradation products and molecules possessing only short poly(A) tracts. Analysis of six cDNA clones generated from the progeny RNA 3 confirmed the presence of a long 3' poly(A) tail in each case but also revealed an unexpected feature: in each case the poly(A) tail was separated from the 3' end of the heteropolymeric portion of the RNA 3 sequence by a short U-rich sequence ranging from one to eighteen nucleotides in length (Fig. 5).

RNA 3		GUAUAUUGAC $(A)_N A_{OH}$
$t35A_0$		———— C_{OH}
PROGENY CLONES	1	———— U $(A)_N A_{OH}$
	2	———— U $(A)_N A_{OH}$
	3	———— UUU $(A)_N A_{OH}$
	4	———— UUAUUUU $(A)_N A_{OH}$
	5	———— UUAUUUUU $(A)_N A_{OH}$
	6	———— UUUUUUUUUUUUUUAUU $(A)_N A_{OH}$

Fig. 5 Sequence at the 3' terminus of the nonpolyadenylated RNA 3 transcript $t35A_0$ and of progeny RNA 3 (as deduced by sequence analysis of six recombinant cDNA clones) from *C. quinoa* local lesions produced by infection with $t35A_0$ plus RNAs 1 and 2.

The fact that multiplication of RNA 3 in the above experiments is always accompanied by the reappearance of a 3' poly(A) tail represents strong circumstantial evidence that the tail is necessary for the replication and/or stability of RNA 3. The enzyme responsible for addition of the tail is at present unknown. A host poly(A) polymerase could be involved or the viral replicase system itself might have such an activity. The significance of the short U-rich tract preceding the newly synthesized poly(A) tail and the mechanism of its synthesis are likewise obscure. Terminal uridylyl transferase activity has been detected in diverse organisms, including plants (Andrews *et al.*, 1985; Zabel *et al.*, 1981; Boege, 1982). It has been suggested that such an activity may intervene in poliovirus replication by adding uridine residues to the 3' poly(A) tail of virion RNA which can then anneal back to the tail and prime minus-strand RNA synthesis (Andrews and Baltimore, 1986). The "aberrant" addition of a U-rich tract to the progeny of the tailless RNA 3 transcript may be a clue that such an activity is involved in normal BNYVV minus-strand RNA synthesis.

The 5' terminal region of RNA 3, or, more strictly speaking, the minus-stand sequence complementary to this region, is the predicted site for a promoter of plus-strand RNA synthesis and is thus also expected to contain sequences essential for RNA 3 accumulation. Of course, sequences required *in cis* for other essential functions such as minus-strand synthesis, RNA stabililty and encapsidation could be located in this region as well. The existence of essential sequences on this portion of the genome was investigated using a nested set of deleted RNA 3 transcripts in which the deletions extended from the AccI site to various restriction sites nearer the 5' terminus (Fig. 3). Accumulation and packaging of these transcripts were unaffected until deletion attained the XmaIII site at nt. 264 at which point and beyond no RNA 3 could be detected. Study of short deletions between neighboring restriction sites near the 5' terminus indicates that the cut-off point for productive replication lies between nt. 265 and nt. 309 (Fig. 3). Deletions eliminating nt. 71-180 and nt. 181-260 were also lethal suggesting that the essential element is a fairly long sequence or secondary structure or consists of a mosaic of separate blocks of sequence. Fine scale deletion mapping and linker mutagenesis will be necessary to delimit the essential elements more precisely. It is noteworthy that, apart from the first 9 residues (Fig. 4), no extensive sequence homology exists between the 5' terminal 300 nt. of RNA 3 and the corresponding regions of RNAs 1, 2 and 4. Thus it will be of great interest to determine if productive replication of the other BNYVV RNAs, for example RNA 4, requires a long 5' sequence.

All of the viable RNA 3 deletions shown in Figure 3 which interrupt or truncate the 25 Kd ORF (i.e., all except 1ΔSE) eliminate the YS lesion phenotype characteristic of the presence of full-length RNA 3. Frameshifts in the 25 Kd ORF caused by linker insertions and conversion of AUG445, the 25 Kd initiation codon, to ACC by oligodeoxyribonucleotide directed mutagenesis also eliminate the YS symptoms (I. Jupin, unpublished observations). Generally, the YS lesions were replaced by pale green spots or rings similar to symptoms produced by RNA 1 plus RNA 2 alone. Two exceptions were noted, however. The deletion variants 24ΔXA and 1ΔEA both produced a severe necrotic local lesion response. The reason why the plant responds differently to these two variants is not understood but, in principle, it should be possible to pinpoint the feature(s) of the viral sequence which elicit the necrotic reaction by construction and analysis of additional variants.

ACKNOWLEDGEMENTS

The authors are indebted to Danièle Scheidecker for excellent technical assistance.

REFERENCES

Albrecht, H., Geldreich, A., Menissier-de Murcia, J., Kirchherr, D., Mesnard, J.M. and Lebeurier, G. (1988) CaMV gene I product detected in a cell-wall-enriched fraction. *Virology* **163**, 503-508.

Ahlquist, P., French, R., Janda, M. and Loesch-Fries, L.S. (1984) Multicomponent RNA plant virus infection derived from cloned viral cDNA. *Proc. Natl. Acad. Sci. USA* **81**, 7066-7070.

Ahlquist, P., French, R. and Bujarski, J.J. (1987) Molecular studies of brome mosaic virus using infectious transcripts from cloned cDNA. *Advan. Virus Res.* **32**, 215-242.

Allison, R.F., Janda, M. and Ahlquist, P. (1988) Infectious *in vitro* transcripts from cowpea chlorotic mottle virus cDNA clones and exchange of individual RNA components with brome mosaic virus. *J. Virol.* **62,** 3581-3588.

Andrews, N.C. and Baltimore, D. (1986) Purification of a terminal uridylyl transferase that acts as host factor in the *in vitro* poliovirus replication reaction. *Proc. Natl. Acad. Sci. USA* **83**, 221-225.

Andrews, N.C., Levin, D. and Baltimore, D. (1985) Poliovirus replicase stimulation by terminal uridylyl transferase. *J. Biol. Chem.* **260**, 7628-7635.

Boege, F. (1982) Simultaneous presence of terminal adenylyl, cytidylyl, guanylyl and uridylyl tranferase in healthy tomato leaf tissue: separation from RNA-dependent RNA polymerase and characterization of the terminal transferases. *Bioscience Rep.* **2**, 379-389.

Bouzoubaa, S., Guilley, H., Jonard, G., Richards, K. and Putz, C. (1985) Nucleotide sequence analysis of RNA-3 and RNA-4 of beet necrotic yellow vein virus, isolates F2 and G1. J. *Gen. Virol.* **66,** 1553-1564.

Bouzoubaa, S., Ziegler, V., Beck, D., Guilley, H., Richards, K. and Jonard, G. (1986) Nucleotide sequence analysis of beet necrotic yellow vein virus RNA-2. *J. Gen. Virol.* **67**, 1689-1700.

Bouzoubaa, S., Quillet, L., Guilley, H., Jonard, G. and Richards, K. (1987) Nucleotide sequence analysis of beet necrotic yellow vein virus RNA-1. J. *Gen. Virol.* **68**, 615-626.

Bouzoubaa, S., Guilley, H. Jonard, G., Jupin, I., Quillet, L., Richards, K., Scheidecker, D. and Ziegler-Graff, V. (1989) Genome organization and function of beet necrotic yellow vein virus. In *"Viruses with Fungal Vectors"* (J.I. Cooper and M.J.C. Asher, eds), AAB, Wellesbourne, UK.

Brunt, A.A. and Richards, K.E. (1989) Biology and molecular biology of furoviruses. *Advan. Virus Res.* **36**, 1-32.

Burgermeister, W., Koenig, R., Weich, H., Sebald, W. and Laseman, D.E. (1986) Diversity of the RNAs in thirteen isolates of beet necrotic yellow vein virus in *Chenopodium quinoa* detected by means of cloned cDNAs. *J. Phytopathol.* **115**, 229-242.

Dawson, W.O., Beck, D.L., Knorr, D.A. and Grantham, G.L. (1986). cDNA cloning of the complete genome of tobacco mosaic virus and production of infectious transcripts. *Proc. Natl. Acad. Sci. USA* **83**, 1832-1836.

Deborde, D.C., Naeve, C.W., Herlocher, M.L. and Maassab, H.F. (1986) Resolution of a common RNA sequencing ambiguity by terminal deoxynucleotidyl transferase. *Anal. Biochem.* **157**, 275-282.

Domier, L.L., Franklin, K.M., Hunt, A.G., Rhoads, R.E. and Shaw, J.G. (1989) Infectious in vitro transcripts from cloned cDNA of a potyvirus, tobacco vein mottling virus. Proc. *Natl. Acad. Sci. USA* **86**, 3509-3513.

Forster, R.L.S., Beven, M.W., Harbison, S.A. and Gardner, R.C. (1988) The complete nucleotide sequence of the potexvirus white clover mosaic virus. *Nucleic Acids Res.* **16**, 290-303.

Godefroy-Colburn, T., Gagey, M.-J., Berna, A. and Stussi-Garaud, C. (1986) A nonstructural protein of alfalfa mosaic virus in the walls of infected tobacco cells. *J. Gen. Virol.* **67**, 2233-2239.

Goldbach, R. (1987) Genome similarities between plant and animal RNA viruses. *Microbiol. Sci.* **4**, 197-202.

Gorbalenya, A.E., Koonin, E.V., Donchenko, A.P. and Blinov, V.M. (1988) A conserved NTP-motif in putative helicases. *Nature* **333**, 22.

Hamilton, W.D.O. and Baulcombe, D.C. (1989) Infectious RNA produced by in vitro transcription of a full-length tobacco rattle virus RNA-1 cDNA. *J. Gen. Virol.* **70**, 963-968.

Heaton, L.A., Carrington, J.C. and Morris, T.J. (1989) Turnip crinkle virus infection from RNA synthesized *in vitro*. *Virology* **170**, 214-218.

Hodgman, T.C. (1988) A new superfamily of replicative proteins. *Nature* **333**, 22-23, 578 erratum.

John, M., Schmidt, J., Wieneke, U., Kondorosi, E., Kondorosi, A. and Schell, J. (1985) Expression of the nodulation gene nod C of *Rhizobium meliloti* in *Escherichia coli* : role of the *nod C* gene product in nodulation. *EMBO J.* **4** , 2425-2430.

Kamer, G. and Argos, P. (1984) Primary structural comparison of RNA-dependent polymerases from plant, animal, and bacterial viruses. *Nucleic Acids Res.* **12** , 7269-7283.

King, A.M.Q. (1988) Preferred sites of recombination in poliovirus RNA: an analysis of 40 intertypic cross-over sequences. *Nucleic Acids Res.* **16**, 11705-11723.

Kirkegaard, K. and Baltimore, D. (1986) The mechanism of RNA recombination in poliovirus. *Cell* **47**, 433-443.

Koenig, R. and Burgermeister, W. (1989) Mechanical inoculation of sugarbeet roots with isolates of beet necrotic yellow vein virus having different RNA compositions. J. *Phytopathol.* **124**, 249-255.

Kuge, S., Saito, I. and Nomoto, A. (1986) Primary structure of poliovirus defective-interfering particle genomes and possible generation mechanisms of the particles *J. Mol. Biol.* **192** , 473-487.

Lazzarini, R.A., Keene, J.D. and Schubert, M. (1981) The origins of defective interfering particles of the negative-strand RNA viruses. *Cell* **26**, 145-154.

Lemaire, O., Merdinoglu, D., Valentin, P., Putz, C., Ziegler-Graff, V., Guilley, H., Jonard, G. and Richards, K. (1988) Effect of beet necrotic yellow vein virus RNA composition on transmission by *Polymyxa betae*. *Virology* **162**, 232-235.

Linstead, P.J., Hills, G.J., Plaskitt, K.A., Wilson, I.G., Harker, C.L. and Maule, A.J. (1988) The subcellular location of the gene I product of cauliflower mosaic virus is consistent with a function associated with virus spread. *J. Gen. Virol.* **69**, 1809-1818.

Meshi, T., Ishikawa, M., Motoyoshi, F., Semba, K. and Okada, Y. (1986) *In vitro* transcription of infectious RNA from full-length cDNAs of tobacco mosaic virus. *Proc. Natl. Acad. Sci. USA* **83**, 5043-5049.

Morozov, S.Y., Lukasheva, L.I., Chernov, B.K., Skryabin, K.G. and Atabekov, J.G. (1987) Nucleotide sequence of the open reading frames adjacent to the coat protein cistron in potato virus X genome. *FEBS Lett.* **213**, 438-442.

Moser, O., Gagey, M.J., Godefroy-Colburn, T., Stussi-Garaud, C., Ellwart-Tschürtz, M., Nitschko, H. and Mundry, K.W. (1988) The fate of the transport protein of tobacco mosaic virus in systemic and hypersensitive tobacco hosts. *J. Gen. Virol.* **69** , 1367-137.

Petty, I.T.D., Hunter, B.G., Wei, N. and Jackson, A.O. (1989) Infectious barley stripe mosaic virus RNA transcribed *in vitro* from full-length genomic cDNA clones. *Virology* **171**, 342-349.

Putz, C., Pinck, M., Fritsch, C. and Pinck, L. (1983) Identification of the 3' and 5' ends of beet necrotic yellow vein virus RNAs: presence of 3'-poly(A) sequences. *FEBS Lett.* **156**, 41-46.

Quillet, L., Guilley, H., Jonard, G. and Richards, K. (1989) *In vitro* synthesis of biologically active beet necrotic yellow vein virus RNA. *Virology* **172**, 293-301.

Romanova, L.I., Blinov, V.M., Tolskaya, E.A., Viktorova, E.G., Kolesnikova, M.S., Guseva, E.A. and Agol, V.I. (1986) The primary structure of cross-over regions of intertypic poliovirus recombinants: a model of recombination between RNA genomes. *Virology* **155** , 202-213.

Shirako, Y. and Brakke, M.K. (1984) Two purified RNAs of soil-borne wheat mosaic virus are needed for infection. *J. Gen. Virol.* **65**, 119-127.
Steven, A.C., Trus, B.L., Putz, C. and Wurtz, M. (1981) The molecular organization of beet necrotic yellow vein virus. *Virology* **113**, 428-438.
Stussi-Garaud, C., Garaud, J.C., Berna, A. and Godefroy-Colburn, T. (1987) In situ location of an alfalfa mosaic virus nonstructural protein in plant cell walls: correlation with virus transport. *J. Gen. Virol.* **68**, 1779-1784.
Tamada, T. (1975) Beet necrotic yellow vein virus. *CMI/AAB Descr. Plant Viruses* 144.
Tomenius, K., Clapham, D. and Meshi, T. (1987) Localization by immunogold cytochemistry of the virus-coded 30 K protein in plasmodesmata of leaves infected with tobacco mosaic virus. *Virology* **160**, 363-371.
Vos, P., Jaegle, M., Wellink, J., Verver, J., Eggen, R., Van Kammen, A. and Goldbach, R. (1988) Infectious RNA transcripts derived from full-length DNA copies of the genomic RNAs of cowpea mosaic virus. *Virology* **165**, 33-41.
Weiland, J.J. and Dreher, T.W. (1989) Infectious TYMV RNA from cloned cDNA: effects *in vitro* and *in vivo* of point substitutions in the initiation codons of two extensively overlapping ORFs. *Nucleic Acids Res.* **17**, 4675-4686.
Zabel, P., Dorssers, L., Wernoss, K. and van Kammen, A. (1981) Terminal uridylyl transferase of *Vigna unquiculata*: purification and characterization of an enzyme catalyzing the addition of a single UMP residue to the 3'-end of an RNA primer. *Nucleic Acids Res.* **9**, 2433-2453.
Ziegler-Graff, V., Bouzoubaa, S., Jupin, I., Guilley, H., Jonard, G. and Richards, K. (1988) Biologically active transcripts of beet necrotic yellow vein virus RNA-3 and -4. *J. Gen. Virol.* **69**, 2347-2357.
Zuker, M. and Stiegler, P. (1981) Optimal computer folding of large RNA sequences using thermodynamics and auxiliary information. *Nucleic Acids Res.* **9**, 133-148.

DISCUSSION OF K. RICHARDS' PRESENTATION

B. Harrison: Does the coat protein readthrough product occur in the virus particles?

K. Richards: We have not been able to detect it using the antiserum directed against the synthetic peptide which corresponds to the C-terminus of this open reading frame but if the readthrough product is undergoing some sort of maturation or C-terminal degradation we would have missed it.

R. Hull: Are the isolates which lack RNA3 and RNA4 transmissible by *Polymyxa*?

K. Richards: We found that isolates which lacked the small RNAs were very poorly transmitted and the few instances of transmission that occurred were accompanied by re-appearance of the full length RNA3 and RNA4. We suspect that there were trace amounts of RNA3 and RNA4 in the original isolates and that they were amplified as a result of the selection pressure during the natural infection process. One of the main reasons we had for making infectious transcripts was to be able to do this sort of experiment in a more stringent way. I suspect both RNA3 and RNA4 are needed for the natural infection process.

B. Harrison: T. Tamada and colleagues have reported that virus isolates containing only RNA1 and RNA2 are transmitted to a very slight extent by *Polymyxa* whereas virus isolates containing RNA1, 2 and 3 are transmitted much more frequently, but when they also contain RNA4 there is a further dramatic increase in transmission, which is then perhaps 10,000 times more frequent than for isolates containing RNA1 and 2 alone.

A. Miller: Evidence is accumulating for the luteoviruses that a similar coat protein readthrough product may be involved in aphid transmission. Could the same apply to beet necrotic yellow vein virus?

K. Richards: I certainly would not rule out that possibility. The structural analogies are fairly obvious but we have no data yet.

P. Palukaitis: Have you tried ligations other than the one you showed at the BstX1 site to produce the full length clone? In other words making partial digestions and putting the products together in another way?

K. Richards: No.

P. Palukaitis: With cucumber mosaic virus RNA1 we had considerable difficulty in making full length molecules. Initially we did not get any full length clones. Then we cloned the 500 nucleotides at the 5' end. In trying to get full length clones, we found that the bacteria would only tolerate ligations in a certain order when increasingly large clones were made, until eventually the whole molecule was present.

C. Atreya: Did you try any non-bacterial systems for cloning your full length transcripts?

K. Richards: No but we tried every strain of bacteria that we could obtain.

J. Hammond: I suspect that the two deletions that result in a necrotic reaction instead of a chlorotic one may allow the protein still to fold into a globular structure but one with different surface properties.

K. Richards: There are other possibilities too. One is the activation of a small open reading frame, which is normally silent and is just downstream of the deletion. The product of this small open reading frame might be toxic or elicit a hypersensitive response.

Concluding Comments and Reflections on Currents Trends in Plant Virology

B. D. Harrison

Scottish Crop Research Institute, Invergowrie, Dundee DD2 5DA, United Kingdom

In making these concluding comments it seems unnecessary for me simply to attempt to summarize the large amount of information given in the preceding papers. Instead I shall make some points about trends in the development of plant virology, and I shall comment on a few matters which have not received much attention elsewhere in the symposium.

VIRAL GENES AND CONTROL SEQUENCES

The work described in this symposium shows that molecular plant virology is continuing to advance at a great pace. Many viral genes are being characterised at the molecular level, and determination of the complete nucleotide sequences of viral genomes has become commonplace. For example, in the geminivirus group, which was not recognised until 1977, complete sequences of the genomic DNA are now available for about a dozen members. Similarly, among other viruses, the time has come when determination of the complete sequences of two viral RNA genomes can constitute merely half the content of a PhD thesis (Ding, 1989).

From a knowledge of viral nucleotide sequences and the recognition of open reading frames, two main technique-led lines of work have developed. One line involves the construction of full-length cDNA clones from which infectious molecules can be derived. This in turn has led to the use of site-directed mutagenesis to study individual viral genes and to the 'reverse genetics' approach to determining gene function. Several papers in this symposium illustrate the great power of these methods.

The second line of work has led via the use of expression systems, to the production in substantial amounts, and characterisation, of individual gene products. These products can be used to make specific antisera which are valuable aids to determining the time course of synthesis, and site(s) of accumulation in cells, of individual viral proteins. In addition, the use of genetic transformation technology to insert individual viral genes into the nuclear genomes of plants is providing further valuable material for experiments on viral gene expression and function.

From the results of all these kinds of experiments an increasingly detailed picture is being put together of how, when and where viral genes function and of how they interact. In addition to information on the roles of coat protein and the viral products involved in nucleic acid replication, much progress is being made in studies on the functions of other non-structural proteins in polyprotein cleavage, virus spread within the plant and vector transmission. For example, the evidence is strengthening for the existence of different mechanisms of cell-to-cell movement in different groups of plant viruses. Whereas tobamoviruses and tobraviruses do not need to produce virus particles to move from cell to cell but require a functional 'movement protein', which associates with plasmodesmata (Tomenius *et al.*, 1987), comoviruses and nepoviruses probably spread from cell to cell as virus particles through virus-induced tubules, which lead to plasmodesmata and contain virus-coded protein (Roberts & Harrison, 1970; Wellink and Van Kammen, 1989 and unpublished results; Harrison *et al.*, 1990). Other viruses may spread by variants of these mechanisms, or in other ways. Another characteristic which is becoming more evident is that a single virus-coded protein may have multiple functions. For example, coat protein not only can assemble with viral nucleic acid to form nucleoprotein particles in which the nucleic acid is protected from inactivators, but probably also has key roles in the initiation of infection (Wilson & Watkins, 1986; Nelson *et al.*, 1987) and, in many instances, in transmission by vectors (Harrison & Murant, 1984).

In parallel with these developments, information is growing on the structures and functions of control sequences, on their interactions with virus-coded proteins and on the interactions between these proteins. Careful analysis of the effects of mutating full length cDNA clones (Ahlquist *et al.*, 1990) seems to be leading to the conclusion that there is little 'padding' in typical plant virus genomes, and that their nucleotide sequences are remarkably tightly adapted to providing an optimum balance of functions.

PLANT VIRUSES AS EVOLVING SYSTEMS

Computer-aided analysis of the primary structure of the proteins they encode has shown that the single-stranded, positive sense RNA genomes of plant viruses can be divided into two main categories, typified by tobamoviruses and comoviruses. These categories have affinities with Sindbis-like and picorna-like viruses of vertebrates, respectively (Goldbach, 1986). Such similarities, and the ability of individual viruses to infect plants and invertebrates, or vertebrates and invertebrates, have led to the speculation that many viruses that infect seed plants or vertebrates may have evolved from viruses of insects (Goldbach, 1986). Such a hypothesis seems to leave open the origin of plant viruses with rod-shaped or filamentous particles. These viruses perhaps have their closest affinities with viruses of fungi or algae, some of which are infected with agents that have morphologically similar particles (Dieleman-Van Zaayen *et al.*, 1970; Gibbs *et al.*, 1975).

The occurrence of RNA recombination between bromovirus genome parts in experimental systems (Bujarski & Kaesberg, 1986), and apparently between homologous genome parts of tobraviruses in nature (Robinson *et al.*, 1987), indicates the existence of an underrated evolutionary mechanism. By such processes whole genes, or blocks of genes, could be acquired from distantly related or unrelated viruses or organisms, leading to evolution by reassortment of modules. Recent sequence information suggests something of this kind may have occurred in the evolution of luteoviruses. Thus whereas the 3' halves of the genomes of three luteoviruses have many similarities, the 5' half of barley yellow dwarf virus RNA resembles that of carmoviruses (Miller *et al.*, 1988) but the 5' half of the RNA of beet western yellows and potato leafroll viruses resembles that of sobemoviruses (Veidt *et al.*, 1988; Mayo *et al.*, 1989; Wilk *et al.*, 1989).

Another possibility is the acquisition of additional genome parts. Thus among furoviruses and allied viruses with plasmodiophoromycete vectors, soil-borne wheat mosaic virus has two genomic RNA segments whereas field isolates of beet necrotic yellow vein virus typically have four RNA species. The two smallest species are not essential for virus replication or fungal transmission but they greatly increase the amounts of virus accumulation in sugar-beet plants and of transmission by *Polymyxa betae* (Lemaire *et al.*, 1988; Tamada *et al.*, 1988). As these species confer advantages on the virus, Jupin *et al.*, (1990) consider them to be genome parts, but one could also make a case for calling them satellite RNA species. Indeed an RNA species of about 0.5×10^6 molecular weight found in some cultures of tomato black ring nepovirus, and which encodes a large protein but has no obvious effect on virus replication or transmission by vectors, is known as a satellite RNA (Murant *et al.*, 1973a; Fritsch *et al.*, 1978). Clearly the distinction between a satellite RNA and a genome part can be almost a matter of semantics, and one may be an evolutionary precursor of the other.

However, despite the increasing interest in evolution by reassortment of RNA modules, other mechanisms are clearly of prime importance, as illustrated by the rapid appearance of new variants of tobacco ringspot virus satellite RNA when artificial mutants are cultured in plants (Gerlach *et al.*, 1990).

SYMPTOM INDUCTION

Despite the greatly increased information now available on molecular aspects of plant virus infections, the processes involved in symptom induction are mostly not well understood. Thus although the development of necrotic lesions is associated with enhanced production of several ‘pathogenesis-related’ proteins (Bol *et al.*, 1990), possibly by a cascade of reactions, the triggering events are not known. Moreover work with the satellite RNA of cucumber mosaic virus shows that symptom production need not always result from the action of a virus-coded protein. When included in virus inocula, variant forms of this satellite RNA can cause the plant to react by developing necrosis, yellow mosaic or inapparent infection. These satellite RNA species seem not to exert their effects through an encoded polypeptide, and their yellow mosaic-inducing and necrosis-inducing domains have been assigned to a few nucleotides near the middle and towards the 3' end, respectively, of the satellite RNA molecule (Devic *et al.*, 1989).

However, it is an oversimplification to suppose that all symptoms of infection result directly from events at the molecular level. Several types of symptom may represent indirect effects. For example, the symptoms associated with occlusion of phloem cells by callose are well known. Similarly, plasmodesmatal permeability may be affected. Infection with carrot mottle virus can result in the formation of long outgrowths from plasmodesmata into the cytoplasm (Murant *et al.*, 1973b), and this probably explains the decrease in cell-to-cell movement of fluorescent peptides that is found in tissue infected with this virus (Derrick *et al.*, 1989). In contrast, expression of the movement protein of tobacco mosaic virus in transgenic plants is accompanied by an increase in the plasmodesmatal size exclusion limit (Wolf *et al.*, 1989). Such changes probably affect physiological processes in the plant and may be involved in symptom development.

VIRUS INTERACTIONS

Although virologists often go to great trouble to work with inocula containing a single virus, multiple infections are common in nature so providing many opportunities for interactions between viruses to occur. For example, the concentration of one of a pair of viruses may be increased as the result of co-infection with the other (Rochow & Ross, 1955), perhaps because the dependent virus can use the RNA replication machinery of the helper virus. In other instances, the helper virus can provide a cell-to-cell movement function, as when infection with potato virus X breaks down the resistance of tomato plants to tobacco mosaic virus (Taliansky *et al.*, 1982) or when co-infection with potyviruses or tobraviruses increases the cell-to-cell spread of potato leafroll virus (Barker, 1989). In a third kind of interaction the helper virus provides the coat protein. For example, carrot red leaf luteovirus provides the coat protein which not only packages the RNA of carrot mottle virus but also thereby enables this dependent virus to be transmitted by the aphid vector of the luteovirus (Waterhouse & Murant, 1983).

Other interactions which affect vector transmission are found when the helper virus provides a helper component which can enable a dependent virus from another taxonomic group to be transmitted. For example, potato Y potyvirus assists the transmission by aphids of potato aucuba mosaic potexvirus in this way (Kassanis & Govier, 1971), and anthriscus yellows virus assists that of parsnip yellow fleck virus (Murant & Goold, 1968). In another type of interaction which is not fully understood, the flexuous particles of heracleum latent virus become attached individually to the ends of the flexuous particles of heracleum virus 6 in co-infected plants. This association enables heracleum latent virus to be transmitted by the aphid vector of the helper virus but does not occur when preparations of particles of the two viruses are mixed *in vitro* (Murant & Duncan, 1984).

Finally, some interactions involve three partners, as recently discovered in work on groundnut rosette, the most important virus disease of groundnut (peanut) in Africa. It transpires that the disease is caused by satellite RNA, which depends on the non-aphid-transmissible groundnut rosette virus for its replication, with both depending on the aphid-transmissible groundnut rosette assistor luteovirus to provide the coat protein in which viral and satellite RNA are packaged, so enabling them too to be aphid-transmitted (Murant *et al.*, 1988).

From these few examples it seems that interactions can occur in which one virus uses almost any gene product of another virus. These interactions can have important effects on symptoms, on virus spread within the plant and on virus transmission by vectors, and they are undoubtedly commoner than is recognised currently.

APPLICATIONS OF MOLECULAR PLANT VIROLOGY

An important justification of research on plant viruses is that it should lead to improved methods of preventing virus disease. The production of genetically transformed plants which express coat protein-mediated resistance (Abel *et al.*, 1986) or satellite-mediated resistance (Gerlach *et al.*, 1987; Harrison *et al.*, 1987) has already produced some spectacular results, and work on plants transformed with ribozymes (Gerlach *et al.*, 1990) shows promise. To provide further options for virus control, more knowledge at the molecular level is needed on the processes which lead to symptom development and on the mechanisms of virus transmission by vectors. These processes will involve interactions between viral materials, between viral and host materials and between viral and vector materials at the nucleic acid/nucleic acid, nucleic acid/protein and protein/protein levels. Evidence now emerging of many of these kinds of interaction has been described in this symposium. However, this is just a small beginning in what is likely to become a key area of research.

A second development is the increasing use, in other spheres, of knowledge gained from research on plant viruses. Plant viruses have been valuable sources not only of promoter and leader sequences for use in plant molecular biology, but also of systems which have had a major influence on the development of ribozymes. I believe that plant viruses will continue to be important in a wider context, as well as for their own sake, because they exemplify relatively small pieces of genetic material which operate in subtle and intricate ways, and which constitute superb experimental material for examining the mechanisms involved in genome structure, function and evolution.

REFERENCES

Abel, P.P., Nelson, R.S., De, B., Hoffman, N., Rogers, S.G., Fraley, R. T. and Beachy, R.N. (1986). Delay of disease development in transgenic plants that express the tobacco mosaic virus coat protein gene. *Science* **232**, 738-743.

Ahlquist, P., Allison, R., Dejong, W., Janda, M., Kroner, P., Pacha, R. and Traynor, P. (1990). Molecular biology of bromovirus replication and host specificity. This volume, p. 144-155.

Barker, H. (1989). Specificity of the effect of sap-transmissible viruses in increasing the accumulation of luteoviruses in co-infected plants. *Ann. Appl. Biol.* **115**, 71-78.

Bol, J. et al. (1990). Induction of host genes by the hypersensitive response of tobacco to virus infection. This Volume, p. 1-12.

Bujarski, J.J. and Kaesberg, P. (1986). Genetic recombination between RNA components of a multipartite plant virus. *Nature*, London **321**, 528-531.

Derrick, P.M., Barker, H. and Oparka, K.J. (1989). Effects of viral infection on intercellular symplastic movement. *Ann. Rep. Scottish Crop Res. Inst.* 1988 (in press).

Devic, M., Jaegle, M. and Baulcombe, D. (1989). Symptom production in tobacco and tomato is determined by two distinct domains of the satellite RNA of cucumber mosaic virus (strain Y). *J. Gen. Virol.* **70**, 2765-2774.

Dieleman-van Zaayen, A., Igesz, O. and Finch, J.T. (1970). Intracellular appearance and some morphological features of viruslike particles in an ascomycete fungus. *Virology* **42**, 534-537.

Ding, S-W (1989). Molecular biology and evolution of tymoviruses, especially onons yellow mosaic and kennedya yellow mosaic viruses. Ph.D. Thesis, Australian National University, Canberra.

Fritsch, C., Mayo, M.A. and Murant, A.F. (1978). Translation of the satellite RNA of tomato black ring virus in vitro and in tobacco protoplasts. *J. Gen. Virol.* **40**, 587-593.

Gerlach, W.L., Llewellyn, D. and Haseloff, J. (1987). Construction of a plant disease resistance gene from the satellite RNA of tobacco ringspot virus. *Nature*, London **328**, 802-805.

Gerlach, W.L., Haseloff, J.P., Young, M.J. and Bruening, G. (1990). Use of plant virus satellite RNA sequences to control gene expression. This Volume, p. 177-186.

Gibbs, A., Skotnicki, A.H., Gardiner, J.E., Walker, E.S. and Hollings, M. (1975). A tobamovirus of a green alga. *Virology* **64**, 571-574.

Goldbach, R.W. (1986). Molecular evolution of plant RNA viruses. *Ann. Rev. Phytopathol.* **24**, 289-310.

Harrison, B.D. and Murant, A.F. (1984). Involvement of virus-coded proteins in transmission of plant viruses by vectors. In Vectors in Virus Biology, ed. M.A. Mayo and K.A. Harrap. London: Academic Press Inc., p. 1-36.

Harrison, B.D., Mayo, M.A. and Baulcombe, D.C. (1987). Virus resistance in transgenic plants that express cucumber mosaic virus satellite RNA. *Nature*, London **328**, 799-802.

Harrison, B.D., Barker, H. and Derrick, P.M. (1990). Intercellular spread of potato leafroll luteovirus: effects of co-infection and plant resistance. In Recognition and Response in Plant Virus Interactions, ed. R.S.S. Fraser. Berlin: Springer, (in press).

Jupin, I., Quillet, L., Niesbach-Klosgen, U., Bouzoubaa, S., Richards, K., Guilley, H. and Jonard, G. (1990). Infectious synthetic transcripts of beet necrotic yellow vein virus RNAs and their use in investigating structure-function relations. This Volume, p. 187-204.

Kassanis, B. and Govier, D.A. (1971). The role of the helper virus in aphid transmission of potato aucuba mosaic virus and potato virus C. *J. Gen. Virol.* **13**, 221-228.

Lemaire, O., Merdinoglu, D., Valentin, P., Putz, C., Ziegler-Graff, V., Guilley, H., Jonard, G. and Richards, K. (1988). Effect of beet necrotic yellow vein virus RNA composition on transmission by Polymyxa betae. *Virology* **162**, 232-235.

Mayo, M.A., Robinson, D.J., Jolly, C.A. and Hyman, L. (1989). Nucleotide sequence of potato leafroll luteovirus RNA. *J. Gen. Virol.* **70**, 1037-1051.

Miller, W.A., Waterhouse, P.M. and Gerlach, W.L. (1988). Sequence and organization of barley yellow dwarf virus genomic RNA. *Nucleic Acids Res.* **16**, 6097-6111.

Murant, A. and Duncan, G. (1984). Nature of the dependence of heracleum latent virus on heracleum virus 6 for transmission by the aphid Cavariella theobaldi. Abstr. Sixth. Int. Cong. Virology, Sendai, Japan, p. 328.

Murant, A.F. and Goold, R.A. (1968). Purification, properties and transmission and parsnip yellow fleck, a semi-persistent aphid-borne virus. *Ann. Appl. Biol.* **62**, 123-137.

Murant, A.F., Mayo, M.A., Harrison, B.D. and Goold, R.A. (1973a). Evidence for two functional RNA species and a `satellite' RNA in tomato black ring virus. *J. Gen. Virol.* **19**, 275-278.

Murant, A.F., Roberts, I.M. and Goold, R.A. (1973b). Cytopathological changes and extractable infectivity in Nicotiana clevelandii leaves infected with carrot mottle virus. *J. Gen. Virol.* **21**, 269-283.

Murant, A.F., Rajeshwari, R., Robinson, D.J. and Raschke, J.H. (1988). A satellite RNA of groundnut rosette virus that is largely responsible for symptoms of groundnut rosette disease. *J. Gen. Virol.* **169**, 1479-1486.

Nelson, R.S., Abel, P.P. and Beachy, R.N. (1987). Lesions and virus accumulation in inoculated transgenic tobacco plants expressing the coat protein gene of tobacco mosaic virus. *Virology* **158**, 126-132.

Roberts, I.M. and Harrison, B.D. (1970). Inclusion bodies and tubular structrues in Chenopodium amaranticolor plants infected with strawberry latent ringspot virus. *J. Gen. Virol.* **7**, 47-54.

Robinson, D.J., Hamilton, W.D.O. Harrison, B.D. and Baulcombe, D.C. (1987). Two anomalous tobravirus isolates: evidence for RNA recombination in nature. *J. Gen. Virol.* **68**, 2551-2561.

Rochow, W.F. and Ross, A.F. (1955). Virus multiplication in plants doubly infected by potato viruses X and Y. *Virology* **1**, 10-27.

Taliansky, M.E., Malyshenko, S.I., Pshennikova, E.S. and Atabekov, J.G. (1982). Plant virus-specific transport function. II. A factor controlling virus host range. *Virology* **122**, 327-331.

Tamada, T., Abe, H., Saito, M., Kiguchi, T. and Harada, T. (1988). The role of beet necrotic yellow vein virus RNA species in symptom expression and fungus transmission. Abstr. 5th Int. Cong. Plant Pathology, Kyoto, Japan, p. 31.

Tomenius, K., Clapham, D. and Meshi, T. (1987). Localization by immunogold cytochemistry of the virus-coded 30K protein in plasmodesmata of leaves infected with tobacco mosaic virus. *Virology* **160**, 363-371.

Veidt, I., Lot, H., Leiser, M., Scheidecker, D., Guilley, H., Richards, K. and Jonard, G. (1988). Nucleotide sequence of beet western yellows virus RNA. *Nucleic Acids Res.* **16**, 9917-9932.

Waterhouse, P.M. and Murant, A.F. (1983). Further evidence on the nature of the dependence of carrot mottle virus on carrot red leaf virus for transmission by aphids. *Ann. Appl. Biol.* **103**, 455-464.

Wellink, J. and Van Kammen, A. (1989). Cell-to-cell transport of cowpea mosaic virus requires both the 58K/48K proteins and the capsid proteins. *J. Gen. Virol.* **70**, 2279-2286.

Wilk, F. van der, Huisman, M.J., Cornelissen, B.J.C., Huttinga, H. and Goldbach, R. (1989). Nucleotide sequence and organziation of potato leafroll virus genomic RNA. *FEBS Letters* **245**, 51-56.

Wilson, T.M.A. and Watkins, P.A.C. (1986). Influence of exogenous viral coat protein on the cotranslational disassembly of tobacco mosaic virus (TMV) particles in vitro. *Virology* **149**, 132-135.

Wolf, S., Deom, C.M., Beachy, R.N. and Lucas, W.J. (1989). Movement protein of tobacco mosaic virus modifies plasmodesmatal size exclusion limit. *Science* **246**, 377-379.

Index of Contributors

Entries in bold type indicate papers, or contribution by session chairman; presenters of papers are indicated by asterisks. Entries not in bold type refer to discussion contributions.